SpringerBriefs in Computer Science

Series Editors

Stan Zdonik
Peng Ning
Shashi Shekhar
Jonathan Katz
Xindong Wu
Lakhmi C. Jain
David Padua
Xuemin Shen
Borko Furht
V. S. Subrahmanian
Martial Hebert
Katsushi Ikeuchi
Bruno Siciliano

For further volumes:
http://www.springer.com/series/10028

Rapid Serial Visual Presentation

rsvp

Robert Spence
Mark Witkowski

Robert Spence · Mark Witkowski

Rapid Serial Visual Presentation

Design for Cognition

 Springer

Robert Spence
Department of Electrical and Electronic
Engineering
Imperial College London
London
UK

Mark Witkowski
Department of Electrical and Electronic
Engineering
Imperial College London
London
UK

ISSN 2191-5768   ISSN 2191-5776   (electronic)
ISBN 978-1-4471-5084-8   ISBN 978-1-4471-5085-5   (eBook)
DOI 10.1007/978-1-4471-5085-5
Springer London Heidelberg New York Dordrecht

Library of Congress Control Number: 2013933580

Printed on acid-free paper

Springer is part of Springer Science+Business Media (www.springer.com)

Preface

The action of rapidly riffling the pages of a book in order to gain some idea of its content is called Rapid Serial Visual Presentation (RSVP). Digital implementation of RSVP permits a wide range of RSVP modes, and has many applications.

While the RSVP of text has been studied extensively, the computational resources required to flexibly achieve the RSVP of images have only effectively been available for around 20 years. Considerable progress has been made since that time both as far as applications are concerned as well as our understanding of RSVP. It is therefore timely to devote a monograph to the RSVP of images.

A conscious decision has been made to choose the interaction designer as the principal user of this monograph, a decision that is reflected by the inclusion of design considerations in many chapters. Other interested readers of the monograph could well include cognitive and perceptual psychologists, since there is much that we do not know about RSVP: especially in Chap. 5, there is a substantial content of previously unpublished research.

Imperial College London, December 2012
Robert Spence
Mark Witkowski

Acknowledgments

It is a distinct pleasure to express our gratitude for the collaboration we have enjoyed over a period of almost 20 years with a number of people involved in research into RSVP. From the late 1990s, when the influence of eye-gaze was first identified, they included Oscar de Bruijn, Kenny Tong and Kevin Lam. The new century saw Catherine Fawcett and Katy Cooper carry out their experiments with a range of RSVP modes. Soon after that we enjoyed a collaboration with Marco Porta, Mauro Mosconi, Alice Ravarelli and Andrea Bianchi of the Laboratorio di Visione Artificiale, Dip. di Informatica e Sistemistica, Università di Pavia, Italy and some of our gaze analysis studies are derived from records generously provided by them. More recently our colleagues Tim Brinded and James Mardell have enhanced our understanding of RSVP. We especially appreciate Kent Wittenburg's longstanding interest in our work, especially since it comes from one who has contributed immensely to the industrial application of RSVP in the important area of consumer and other products.

Robert Spence gratefully acknowledges the support provided between 2000 and 2003 by the UK's Engineering and Physical Sciences Research Council (EPSRC) for a project entitled The Usefulness and Usability of Space–time Trade-offs for Information Navigation on Small Displays. He also benefited from the award of an Emeritus Fellowship (2010–2012) from the Leverhulme Trust.

Contents

Contents

Chapter 1
What is RSVP? And Why do I Need it?

Abstract The topic of Rapid Serial Visual Presentation (RSVP) is introduced using a very familiar example, that of rapidly riffling the pages of a book in order to locate a needed image. Advances in computation and graphical processing now enable the benefit of RSVP to be exploited in a wide variety of common tasks. The fifteen illustrative examples provided in this chapter include window-shopping, video fast forward and rewind, searching on a mobile device and searching for information under pressure.

Keywords Rapid serial visual presentation • Pre-attentive processing • Applications • Interaction design • Product browsing • Product selection • Gist • Video search and editing

1.1 Bookshops

You walk into a bookshop and notice, among all the books on display, one whose cover design or title suggests that it might be of interest to you. So you pick it up, flex it somewhat and then riffle fairly quickly through its pages to get a first glimpse of its contents (Fig. 1.1). Are there many pictures? What about equations? Probably in the course of four or five seconds you will know whether further, and more detailed, inspection is warranted. Because the action of riffling is essentially a rapid one, and the pages are viewed in sequence, we say that you are experiencing a Rapid Serial Visual Presentation of the pages or, for short, RSVP.

Nowadays, of course, both computation and graphical processing are developed to such a degree that this means of presentation can be achieved by computer. What is more, many variants of RSVP can be contemplated that would be difficult or impossible to achieve by physical means. As a consequence, RSVP can support a wide variety of common and not-so-common tasks in a wide range of domains, as examples in this chapter will illustrate (Spence 2002). First, though, it is useful briefly to identify certain aspects of RSVP that make it so attractive.

Fig. 1.1 Fast riffling of the pages of a book either to gain an appreciation for its content or to locate a wanted image

1.2 Text RSVP

It is important here to point out what this book is NOT about. An associated field of investigation, which is specifically not addressed in this book, is the Rapid Serial Presentation of Text in which individual words (or sometimes multiple words) are presented sequentially in a fixed place on a display. 'Text RSVP' has been extensively researched and finds application in speed-reading, information presentation for small displays, assisted reading for the disabled and as a diagnostic tool for dyslexia.

Studies comparing text mode RSVP with standard page-based text presentation indicated that substantial reading speed gains can be achieved with little deterioration of measured comprehension (e.g. Rubin and Turano 1992). They conclude that these performance gains are due to the reduction of saccadic eye movements, and in that respect share some concepts with Rapid Serial *Visual* Presentation, which is the subject of this book. A large number of speed-reading applications[1] have become available and the technique finds obvious application on mobile devices with limited display resolution and size.[2]

---

[1] For instance: http://www.speedreadingblogger.com/website-of-the-day/productivity-tools/how-to-read-faster-on-the-computer-screen-using-rapid-serial-visual-presentation-rsvp/.

[2] For instance: http://wiki.forum.nokia.com/index.php/Information_Visualization:_Rapid_Serial_Visual_Presentation_(RSVP).

1.3 Images

This book's exclusive focus on images that are presented in rapid succession is exemplified by the riffling of the pages of the book you picked up in the bookshop: you probably saw each page for about one-fifth to one-tenth of a second. Nevertheless, if your interest was in the History of Art, you would immediately recognise the Mona Lisa or one of Mondrian's distinctive paintings well within that short exposure. That's a very short time in which to do something useful! What's more, you didn't have to pay any conscious *attention* in order to identify those images of interest. That is why we refer to *pre-attentive visual processing*, of which more later. Therefore, two good reasons for exploring the potential offered by RSVP are that something of interest (either detailed or generic) can be identified *very quickly* and *without conscious cognitive effort*. For those two simple but profound reasons the potential of RSVP certainly deserves exploration. First, though, we need to ask a question about its use.

1.4 Who Needs RSVP?

Just about everyone needs RSVP! The relevance of computer-supported RSVP to common and not-so-common human activities is so extensive that it is helpful first to examine a number of different illustrative applications, not only to be convinced about the widespread relevance of RSVP but additionally to establish the sort of tasks that can be supported. The applications described below also serve to illustrate some of the many possible presentational 'modes' of RSVP (Porta 2009).

1.4.1 Example 1: Where's That Photo?

You have a collection of digital photos—probably an extensive collection—and you wish to find a specific one to show to a friend. If your laptop were to present that collection in sequence, with one image replacing another (Fig. 1.2) and with each photo presented for only 100 ms, there is an extremely high probability—around 95 %—that you would identify the required photo and then be able to retrieve it. In this way an efficient search can be carried out at a rate of ten photos per second, or 600 in a minute.

The presentational technique illustrated in Fig. 1.2 is called Slide-show RSVP mode. As well as supporting the task of identifying one specific 'target' image within a collection it can also support a search for a generic target: you may have some photos of vintage railway engines and merely want to select a representative one.

The so-called Collage mode RSVP (Wittenburg et al. 2000, 1998) illustrated in Fig. 1.3 for the activity of choosing a book to buy has something in common with the Slide-show mode: it simulates the placing of book covers on a table at a rate of about one every 200 ms. An advantage over the Slide-show mode is that images remain visible until they are covered by later images. A manual control (see arrows at the lower edge of the display) allows the rate of presentation to be varied and reversed.

Fig. 1.2 Slide-show RSVP:
each image occupies the
whole viewing area for a
short time

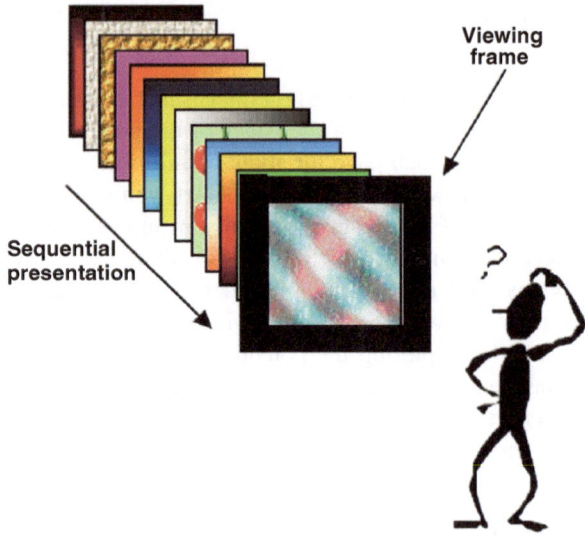

Fig. 1.3 Collage mode
RSVP: analogous to books
being thrown onto a table at
a rate of about 5 per second.
Overlap may well occur
(courtesy of Kent Wittenburg)

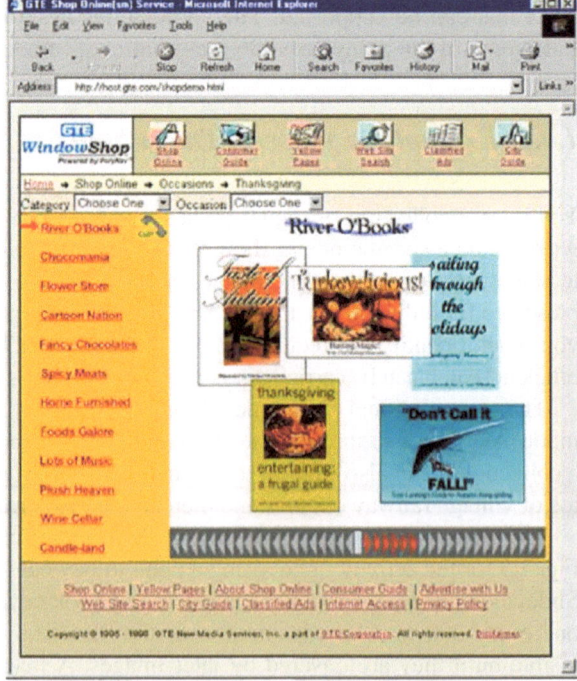

1.4.2 Example 2: What's on the TV?

You return from a hard day at work, grab a drink, sit down and decide to watch television. But which of the many available channels? One approach, and one imposed by earlier television sets, is to flick through the channels, a tedious procedure and somewhat unattractive if there are hundreds available. As an alternative a different mode of RSVP can help (Fig. 1.4) by presenting a moving cascade of images, each of which is the current frame of a different TV channel (Wittenburg et al. 2003b, 2000) but with one magnified and 'captured' to remain static for a short period (e.g., 200 ms). Additionally, the images are sized to provide a 3D effect, which users may find attractive.

With this mode it is possible to identify a potentially interesting channel in a quick glance (e.g., one-fifth of a second) so that the speed of image movement could be quite high (how high we shall see later). Metadata in the form of the channel identification ('Channel 5') is provided. With such an interface a suitable channel may be identified well within a minute. We note that what is usually being sought in this example is a single channel but often a choice that 'satisfices' rather than 'optimises'.

1.4.3 Example 3: Oops—It's Mothers' Day!

It's Mothers' Day tomorrow, you've forgotten to buy her a present and the local department store is closed: online shopping is indicated. But you have no idea what to buy, so the sequential examination of many websites becomes tedious and

Fig. 1.4 A continuously moving cascade of images, each of which is the current frame of a different TV channel, allows rapid browsing of available channels and the selection of one to view

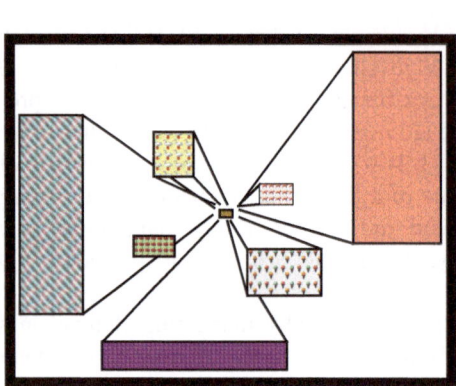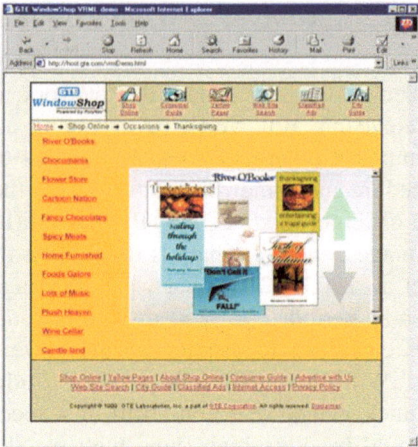

Fig. 1.5 Floating mode RSVP in which images appear to approach the viewer from a distance: diagrammatic representation (*left*), an implementation—*sensitive arrows* allow the speed and direction of 'movement' to be controlled by the user (courtesy of Kent Wittenburg) (*right*)

time-consuming. Instead, the Floating mode of RSVP represented diagrammatically in Fig. 1.5 (left) can, in a few seconds, present images associated with a wide range of possible presents: the images appear to 'move' towards the viewer much like the billboards beside a motorway 'move' past a car (Fig. 1.5, right). However, when a possible present is noticed the movement can be reversed by the user to return that item to view, after which a click on the image causes a full-screen description to be presented (Wittenburg et al. 2000).

The Floating mode of RSVP is especially useful in those situations where a target has only been formulated in general terms, and where an outcome that *satisfices* is perfectly acceptable. We acknowledge the fact that in the floating mode images are presented in parallel as well as serially. It is useful to observe that, with images, only a small size of image—as at the 'start' of its trajectory—may be sufficient for its relevance to be judged: the time that an image then spends on the display may help the user to confirm the relevance of that image.

1.4.4 Example 4: Choosing Music or a Book

An attractive way of browsing through a collection of music covers or books in order to select one to examine in more detail is to employ the CoverFlow mode of RSVP (see US patent D613,300, April 2010) (Fig. 1.6). The speed with which the different covers move in and out of view can be high and warrant the term 'RSVP', but manual control provides a convenient means of controlling the rate of presentation in order to identify and select a particular item of interest.

Fig. 1.6 The CoverFlow mode of RSVP, with convenient manual control of speed

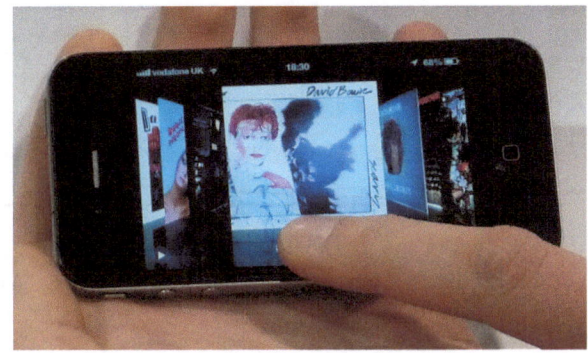

Fig. 1.7 In a 'paperless' business meeting, rapid retrieval of a pertinent page is often required

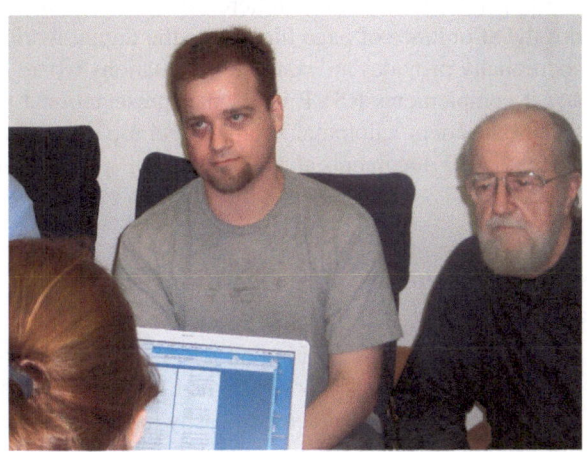

1.4.5 Example 5: Professional Meetings

Professional meetings in business and the law courts, to select just two examples, are increasingly characterised by an absence of paper, with most participants viewing documents on a display. The young lady in Fig. 1.7 is in negotiation with two gentlemen whose time is valuable, so she wants to find, as quickly as possible, the page in her document that illustrates the point she is making. With conventional documents one has to move from one page to the next until the relevant page is identified: less time-consuming and stressful would be a Slide-show RSVP of the document's pages, especially if the user can recall and recognise the nature of the required page (e.g., "Green and red diagram upper left, little text, table bottom right"). Here the user is *searching* for one familiar target when only that specific target is acceptable.

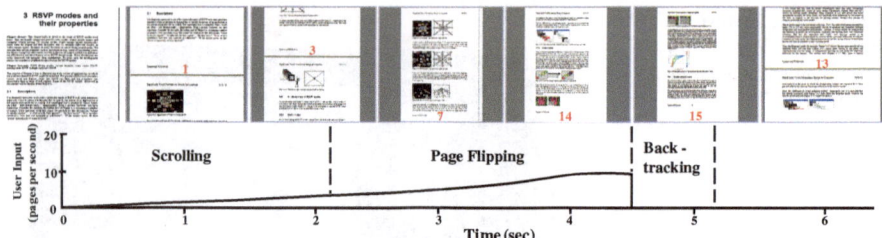

Fig. 1.8 In finding one's way through a document, a speed-dependent mixture of slide-show RSVP and scrolling may facilitate search or browsing (adapted from Sun and Guimbretiere (2005))

For such an application an alternative to Slide-show RSVP is available in the form of Flipper (Sun and Guimbretiere 2005) which, in the words of its creators, "carries the affordance of page flipping to the digital world" (Fig. 1.8). Flipper, however, additionally provides an example of situations where RSVP is *not* always appropriate: it complements RSVP with other presentational techniques. "Flipper combines speed-dependent automatic zooming and rapid serial visual presentation to let users navigate their documents at a wide range of speeds" (Sun and Guimbretiere 2005).

1.4.6 Example 6: The Graphic Designer

A graphic designer is creating an advertisement for a client and requires a 'palette' of about ten images from which one or two can be selected for inclusion in the advertisement. The designer may easily have a huge number of images to draw upon (for example from an agency's database), so it is not essential to locate *every* image that might be relevant: a sample of ten may be sufficient to examine more carefully before a final choice is made. Here, the RSVP Volcano mode illustrated in Fig. 1.9 is one possibility: images emerge from a central location and then 'flow' outwards along one of many paths, decreasing in size as they move (Porta 2006, 2009).

1.4.7 Example 7: The Uninformative Folder

I have a digital folder on my desktop with the ill-chosen title "New" (Fig. 1.10a). Even though I created and filled that folder I cannot now remember its content: the question "What's in there?" is triggered. The conventional approach of opening the folder and then examining the separate files in turn is laborious. More attractive by far is the prospect of being able to riffle through its contents. Figure 1.10b shows an early proposed alternative (Spence 1998) suited to collections of images: interaction causes the images to emerge at recognisable size from the left of the

Fig. 1.9 In Volcano mode RSVP images 'emerge' at a *central location* and then 'flow' towards the *outer edges*, becoming smaller as they travel

(a)   **(b)**   **(c)**

Fig. 1.10 "What's in there?" A slide-show RSVP of the folder's contents would help to answer this question quickly

folder, execute a roughly circular path and then re-enter the folder on the right. The fraction of the image collection currently displayed is indicated on the folder. For reasons that will become clear in Chap. 3 this technique would benefit if, in at least one image location (probably top-dead centre), an image were stationary for about 100 ms. The use of riffling to browse an image collection is, of course, present in the 'Photo' app on an iPad (Fig. 1.10c).

1.4.8 Example 8: Gist of a Television Programme

I look at a schedule of TV programmes and notice a film called 'Flame'. Do I want to watch it? In the absence of a published review (which I'd have to read) I have no idea of the film's genre. One solution to the problem of acquiring the 'gist' of a film is to present, in Slide-show mode RSVP (Fig. 1.11) a sequence of 'key frames' specifically extracted (by a specialist) from the film to illustrate its nature. Experiments have shown (Tse et al. 1998) that, within about 2 s, 20 appropriately

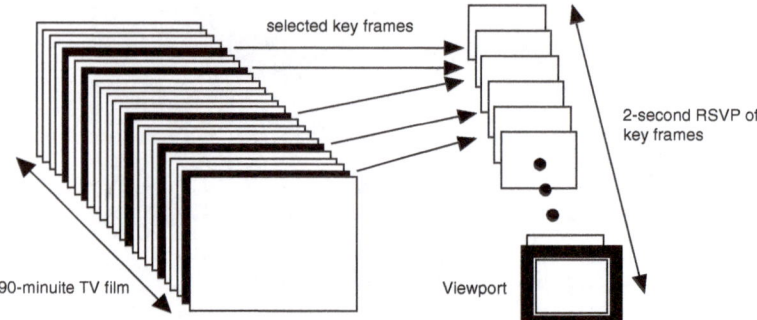

Fig. 1.11 A rapid slide-show presentation of key frames of a film can quickly convey the gist of that film

Fig. 1.12 Layout of the SeeHere interface

selected key frames are sufficient to inform the viewer about the gist of a film. Indeed, modern film trailers presented on TV channels provide a good example of a rapid sequence of key frames carefully chosen to convey gist, although the time for which each frame is visible is somewhat longer than 100 ms.

Using the same principle of the presentation of key frames, Wittenburg et al. (1999, 2000) developed SeeHere, another design that supports rapid and efficient acquisition of the gist of each of a collection of videos. For each video currently of interest a transparent title appears in the centre of the display (Fig. 1.12), while eight representative key frames appear one after another in the lower left quadrant and jump, in a predictable temporal sequence, clockwise from one quadrant to another.

The positioning of each image has an element of randomness "to interject a note of serendipity and visual appeal". The dashed yellow squares in Fig. 1.12 have been added artificially to explain the rule governing image position. When an image is in (say) the top-left part of the display, the top-left corner of the image is randomly positioned in the (invisible) top-left yellow square. This degree of random positioning also facilitates the use of images of different sizes and aspect ratios. Each new group of key frame images associated with a new video title would be separated from the previous by an empty background. Wittenburg et al. (2000) notes "when compared to Slidemode presentations, each image is visible longer, and a set of images visible at the same time can reinforce a particular cinematographic style".

On the basis of experience the developers pointed to the effective integration of temporal and spatial layout embodied in SeeHere and hypothesised that, for each video title, eight key frames displayed at four frames per second would give a viewer sufficient opportunity to acquire a high-level gist of a video title, thereby supporting the browsing of about thirty titles per minute.

1.4.9 Example 9: The Mobile Keyhole Problem

Using the miniscule display of a mobile telephone to explore the vastness of the Web is a daunting challenge, but one that RSVP can facilitate to a useful extent. Take the example of a Web page associated with news (as illustrate by the sketch of Fig. 1.13a). Typically the main area is composed of a number of modules, each containing an easily recognised representational image and a few key words. Alternatively (Fig. 1.13b) each module can be presented—*at roughly the same size*—on the display of a mobile or PDA for about half a second, and followed by other modules, the entire presentation taking about 5 s. We could say here that RSVP offers a 'space-time trade-off' (de Bruijn and Tong 2003).

1.4.10 Example 10: Photos on a Mobile

A Nokia mobile phone design allows the owner to view and select from a large collection of photographic images. This facility, designed by Ron Bird, uses the Flip Zoom mode of RSVP (Fig. 1.14) which displays a small number of the images and permits scrolling in a circular movement, with the convenience that one visible image is shown magnified (Holmquist 1997).

1.4.11 Example 11: Selecting a Video

Lam and Spence (1997) reported a technique for the selection of a video from a large collection (Fig. 1.15). Each video is represented by a poster, and those posters are 'stacked' to form a three-tier 'bookshelf'. Cursor movement along a stack causes posters to briefly 'pop out' sideways, allowing them to be examined at a rate of up to 10 per second depending on the speed of cursor movement. The

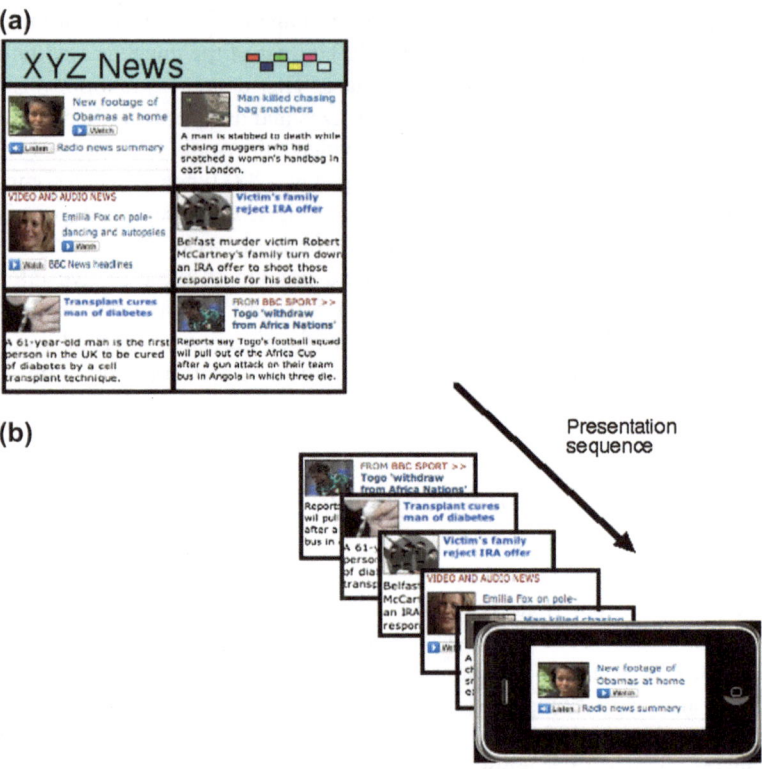

Fig. 1.13 **a** a news web page; **b** segmentation of the page and continuous presentation, in Slide-show mode, on a mobile, to provide a user with a quick appreciation of the topic of available news stories

Fig. 1.14 Photographs stored on a mobile can be moved around the display at speed, with one centrally located photo enlarged for easier recognition (from Spence 2007, courtesy of Ron Bird)

arrangement of shelves in a bifocal structure (Spence and Apperley 1982) allows posters of interest to be moved towards the central region where they are seen in full such that a short clip or trailer for the video can be played.

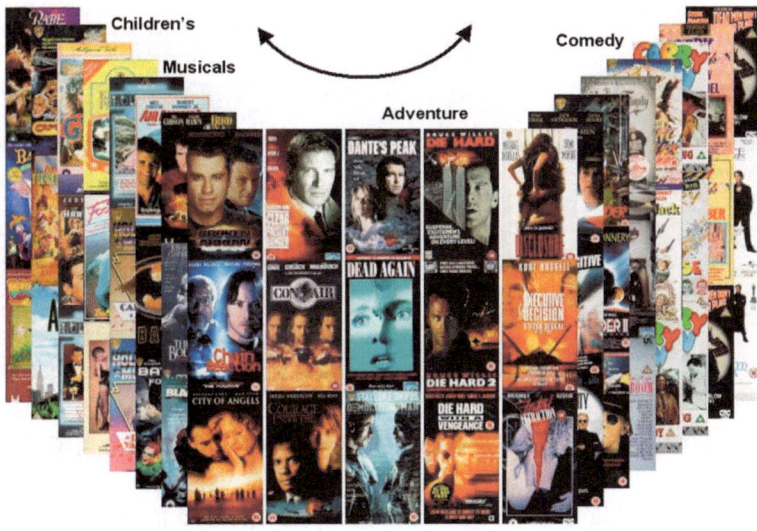

Fig. 1.15 A video store: movement of a cursor along the 'walls' causes posters to pop out. The bifocal arrangement of the wall is such that a click on a full view causes a short movie illustrating the video to be shown (reprinted from Spence 2007)

1.4.12 Example 12: Video Fast-Forward and Rewind

Operations frequently executed by those who record television programmes for later viewing are those of fast-forwarding and rewind. The motivation can vary: from avoiding a commercial break to revisiting a favoured section. Conventionally, controls will support a Slide-show presentation which, though able to be rapid, nevertheless provides no context—one cannot see an approaching break, for example, or a change of scene.

To overcome these limitations Wittenburg et al. (2003b) proposed a variant of RSVP—termed 'Timeshuttle'—which provides valuable context and the consequent opportunity to see where scene changes occur (Fig. 1.16). Manual controls are available to vary both the speed and direction of image movement. Figure 1.16 provides an example of how considerable obscurement of neighbouring images, due to overlap, need not be detrimental (especially when scene changes are of interest), and also serves as a reminder that the interaction designer enjoys considerable flexibility of choice regarding the trajectory followed by a collection of images.

Note that the combination of overlap between successive images and the reducing size of those images together gives rise to a very pronounced 3D depth effect. Wittenburg et al. (2003a) have proposed that this would be effective for data visualization. For example, each image slice might represent geographic data (e.g. rainfall) for a specific year.

Fig. 1.16 An interface designed to facilitate fast-forwarding and rewinding by providing context. The speed of movement can be controlled manually (courtesy of Kent Wittenburg)

1.4.13 Example 13: Purchase Mediated by a Software Agent

Making purchases via the internet is commonplace. Not so common is the experience of receiving advice about your purchase from a personal software agent, externalised on-screen as an animated human-like character capable of speech. An interface illustrating this concept in the context of making a decision about the purchase of wine is shown in Fig. 1.17 (Witkowski et al. 2003). Two key components of the MAPPA interface are of note: a 'shelf' of wine bottles, and 'James', the animated agent whose function is to advise the person purchasing wine.

Fig. 1.17 Agent mediated product browsing. The user can rapidly scan along the whole row of 400 products with a manual control and select ones for description or purchase (reprinted from Witkowski et al. 2003)

The user can view the whole range of wines at leisure by means of a stepping control or can scan back and forth by using a slider control, thereby engaging in a manually controlled RSVP activity. Then, on finding a part of the shelf of particular interest, the user may run through the products by passing the mouse cursor over them. When a bottle is touched it 'pops-up' to be slightly clear of adjacent bottles and a brief 'product label' appears immediately above it. A single-click on a product image triggers the appearance of the agent "James", which offers spoken advice based partly on known user preference. A double-click on a bottle results in a purchase basket dialog box appearing.

Here we have an example where RSVP is not the sole—or even the most crucial—part of an application, and must therefore be considered in context during the design process.

1.4.14 Example 14: Wilderness Search and Rescue

If you become lost in extensive and inaccessible terrain it is increasingly typical for an Unmanned Aerial Vehicle (UAV) to be flown over the area where you are most likely to be found, its video camera transmitting a moving image to a ground station for inspection by a trained human observer. The crucial question concerns the most effective way of carrying out such a visual inspection—that is, to maximise the chance that you will be found. Many approaches are possible. One, of course, is for the human observer simply to view, in real time, the moving image transmitted by the UAV's camera.

Another proposed approach, and relevant to this book, is to break up the terrain image captured by the camera into small segments, then to display each segment statically but sufficiently quickly that the over-flown area is still examined in real time. If the segments are a very small fraction of the overflown area this constraint may mean that each segment is displayed for as short a time as 100 ms. Based on a recent experiment (Mardell et al. 2012) Fig. 1.18 illustrates six possible degrees of segmentation, the highest degree being associated with a display time per segment of 108 ms.

The expectation is that, with such short display times, the human operator's inspection will be pre-attentive rather than involving a conventional visual inspection where the eye is moved around a scene. Surprisingly (especially after seeing such an experiment being conducted), the success in identifying a lost person differed little between a segmentation degree of unity (with about 4 s to inspect each tile) and a segmentation degree of 6 where each map segment is displayed for only 108 ms.

There have been a number of suggested applications using similar segmentation techniques. For instance Forlines and Balakrishnan (2009) proposed segmenting cell-slide pathology preparations, presenting individual cells from the full microscope field in rapid succession. They also suggested that a similar technique might be applied to airport baggage security screening tasks: we will consider this further in Sect. 6.8.

Fig. 1.18 Degrees of segmentation (1–6), corresponding to RSVP slide-show presentation rates in the Search and Rescue task (reprinted from Mardell et al. 2012)

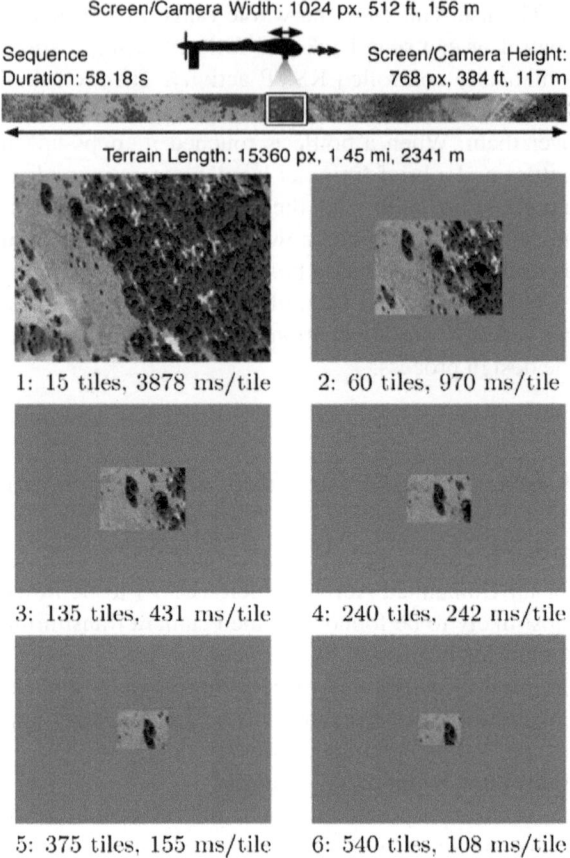

1.5 Tasks: Exploring and Searching

The examples provided above illustrate the wide variety of tasks for which an RSVP can be supportive. Although multiple tasks may be performed while using one particular interface it is nevertheless useful to identify and discuss them.

The tasks can perhaps most conveniently be viewed as falling into two general categories. One is an *exploratory* or *investigative* task, where a user is examining an image collection as though asking the question "what's there?"; TV channel browsing (Example 2), department store exploration (Example 3), film gist acquisition (Example 8), the 'uninformative folder' (Example 7), news selection on a mobile (Example 9), video fast-forward and rewind (Example 12), product purchase (Example 13) and Search and Rescue (Example 14) all provide examples of situations in which the user is forming a mental model of content.

The other type of task involves *search*; the search may be for a specific page in a report (Example 5), a known photograph in a collection (Example 1) or for a small selection of thematic images (Example 6). Sometimes the aim of a search may merely be to satisfy requirements whereas, at other times, a specific 'optimum' may be sought. In many applications, in fact, the user may well switch between exploration and search as a problem becomes better defined, such as with the photographs on a mobile (Example 10) and when exploring and/or searching a video store (Example 11). In such a situation the final design will have to support both component tasks and allow easy (cognitive and interactive) transition between them.

1.6 Questions

Provided with examples that illustrate the many applications of RSVP, and with an appreciation of the tasks that an RSVP might be able to support, the interaction designer will want to know how to proceed with design.

The success of RSVP is intimately associated with properties of the human visual processing system, so it is appropriate to discuss the experimental basis on which these properties are based. To this end we review, in Chap. 2, seven experiments carried out by cognitive psychologists. They begin to establish considerations that are relevant to the RSVP design process.

There are a bewilderingly large number of RSVP modes, only some of which have been used for illustration in this chapter. Which are preferred? And for what reason? Indeed, how should one mode be compared with another? Can any general conclusions be drawn that would begin to help an interaction designer charged with designing any of the systems described above? Chapter 3 aims to address these questions. It first introduces a notation appropriate to the description of RSVP modes and then discusses the properties associated with three main classes of RSVP mode. Again, the outcome identifies considerations relevant to the design of an interactive RSVP application.

Many of the studies reported in Chap. 3 suggest that eye-gaze behaviour may be of profound relevance to the success of any RSVP application. This is indeed the case, so Chap. 4 introduces the phenomenon of eye-gaze and Chap. 5 describes experiments indicating that the effectiveness of an RSVP mode, as well as its acceptance by human users, is closely related to the eye-gaze behaviour of the human user. The content of this latter chapter, much of which is new and previously unpublished, is relevant to the successful design of an RSVP application.

The final Chap. 6 draws upon the outcome of earlier discussions and addresses the questions that would be asked by an interaction designer intending to employ the RSVP technique within an application. It draws upon, but doesn't require knowledge of, the material of Chap. 2, 4 and 5, and might therefore 'stand alone' for the reader already familiar with the potential of RSVP. Chapter 6 concludes with a review of some RSVP applications and suggests unexplored potential.

References

de Bruijn, O., & Tong, C. H. (2003). M-RSVP: Mobile web browsing on a PDA. In E. O'Neill, P. Palanque, & P. Johnson (Eds.), *People and computers: Designing for society* (pp. 297–311). London: Springer.

Forlines, C., & Balakrishnan, R. (2009). Improving visual search with image segmentation. In ACM proceedings of Computer Human Interaction (CHI-09) (pp. 1093–1102). Boston, MA, USA.

Holmquist, L.E. (1997). Focus + Context visualization with Flip-zooming and Zoom Browser, Exhibit, CHI'97.

Lam, K., & Spence, R. (1997). Image browsing: A space-time trade-off (pp. 611–612). *Proceedings INTERACT*. London: Chapman and Hall.

Mardell, J., Witkowski, M., & Spence, R. (2012). An interface for visual inspection based on image segmentation (pp. 697–700). *Proceedings of Working Conference on Advanced Visual Interfaces (AVI-12), Capri Island (Naples)*. Italy: ACM.

Porta, M. (2006). Browsing large collections of images through unconventional visualization techniques (pp. 440–444). *ACM, Proceedings AVI*.

Porta, M. (2009). New visualization modes for effective image presentation. *International Journal of Image Graphics, 9–1*, 27–49.

Rubin, G. S., & Turano, K. (1992). Reading without saccadic eye movements. *Vision Research, 32–5*, 895–902.

Spence, R. (1998). A content explorer. Information Engineering Report 98/08. London: Department of Electrical and Electronic Engineering, Imperial College.

Spence, R. (2002). Rapid, serial and visual: A presentation technique with potential. *Information Visualization, 1*(1), 13–19.

Spence, R. (2007). *Information visualization: Design for interaction*. Englewood Cliffs: Prentice Hall.

Spence, R., & Apperley, M. D. (1982). Data base navigation: An office environment for the professional. *Behaviour and Information Technology, 1*(1), 43–54.

Sun, L., & Guimbretiere, F. (2005). Flipper: A new method for digital document navigation (pp. 2001–2004). *ACM Proceedings CHI'05* (Extended Abstracts).

Tse, T., Marchionini, G., Ding, W., Slaughter, L., & Komlodi, A. (1998). Dynamic Key-frame presentation techniques for augmented video browsing (pp. 185–194). *ACM, Proceedings of Conference on AVI*.

Witkowski, M., Neville, B., & Pitt, J. (2003). Agent mediated retailing in the connected local community. *Interacting with Computers, 15*, 5–32.

Wittenburg, K., Ali-Ahmad, W., LaLiberte, D., & Lanning, T. (1998). Rapid-fire image previews for information navigation (pp. 76–82). *ACM, Proceedings of Conference on AVI*.

Wittenburg, K., Nicol, J., Paschetto, J., & Martin, C. (1999). Browsing with dynamic key frame collages in web-based entertainment video services (Vol. 2, pp. 913–918). *IEEE Proceedings of the International Conference on Multimedia Computing and Systems*.

Wittenburg, K., Chiyoda, C., Heinrichs, M., & Lanning, T. (2000, Jan 26–28). Browsing through rapid-fire imaging: Requirements and industry initiatives (pp. 48–56). *Proceedings of Electronic Imaging '2000: Internet Imaging*. San Jose, CA, USA.

Wittenburg, K., Lanning, T., Forlines, C and Esenther, A. (2003a, June). Rapid serial visual presentation techniques for visualizing a third data dimension. *Proceedings of HCI International Conference*, Crete.

Wittenburg, K., Forlines, C., Lanning, T., Esenther, A., Harada, S., & Miyachi, T. (2003b). Rapid serial visual presentation techniques for consumer digital video devices (pp. 115–124). *ACM, Proceedings Symposium on User Interface Software and Technology (UIST-03)*.

Chapter 2
Experimental Evidence

Abstract Certain features of the human visual processing system influence the success with which the technique of RSVP can be applied. First and foremost is the phenomenon of pre-attentive processing, supporting the recognition of a target image within about 100 ms and without conscious cognitive effort. But other factors, if ignored, can detract from the benefits of RSVP. They include change blindness and saccadic blindness. Other features of the human cognitive system that must be taken into account in any design involving RSVP include user memory and the concept of salience influencing how attention is directed.

Keywords Rapid serial visual presentation • Experimental evidence • Pre-attentive processing • Attention • Brief image presentation • Memory • Attentional blink • Saccadic blindness • Change blindness • Salience

2.1 Psychological Underpinnings of RSVP

In this chapter we describe a number of important experiments, all related to the psychological underpinnings of RSVP in the sense that an understanding of the experimental results enables us to explain observed features of various RSVP applications. That is essential if we are to be able to design RSVPs for a particular purpose. All these experiments illustrate aspects of the human visual processing system; intriguingly, some of the results are remarkably surprising and non-intuitive.

The first two experiments are directly relevant to Slide-show RSVP. Here (Fig. 2.1), a collection of images is shown, one replacing the next at a high rate (typically 10 per second): a subject is asked to say whether a previously seen 'target' image is in that collection or not. The experiments reveal the importance, and some properties, of a very fundamental aspect of human visual processing called **pre-attentive processing**. Basically, pre-attentive processing allows a target image to be recognised within 100 ms without any conscious cognitive effort, a truly attractive feature.

The third and fourth experiments are concerned with a user's memory for one or more rapidly presented images. These two experiments are discussed because,

Fig. 2.1   The role of pre-attentive processing

otherwise, incorrect conclusions may easily be drawn (and, even worse, acted upon) from a cursory glance at the first and second experiments. For example (Fig. 2.5), if the subject is *first* shown a rapid sequence of images and *then* asked "was *this* image present?" the success rate may be very low. In case, in turn, the conclusion is drawn that an exposure of 100 ms is insufficient for an image to be remembered, the fourth experiment demonstrates, surprisingly, that this is not the case.

The last three of the seven experiments demonstrate that a user can be 'blind' in some respect when one image is replaced by another. Again, failure to be aware of these experimental results can be detrimental to the effectiveness of an RSVP application.

2.2 The Recognition of a Briefly Visible Image

One of the earliest experiments relevant to RSVP (Potter and Levy 1969: see also Potter 1975, 1976; Intraub 1981; Coltheart 1999) is summarised in Fig. 2.1. In this simple experiment a subject is first shown a 'target' image and allowed to look at it for a few seconds. It could, for example, be a picture of a briefcase, the person's mother, a car or a painting. The subject is then told that they will see a collection of images presented one after the other at a fast rate, and that their task is to say whether the target image was present in that collection or not.

We illustrate the Slide-show presentation mode by Fig. 2.2 to emphasise that each image occupies the same display area and position: each image appears for (say) 100 ms and is then immediately replaced by the next one in the sequence. Thus, once an image is displayed it does not move: it simply disappears from view after a short time, to be replaced by the next image.

Even at a presentation rate of 10 per second it is found that recognition accuracy can be as high as 90 %. If the images are very distinct it is even possible to achieve reasonable recognition success at a presentation rate of about 20 images per second. Clearly, the human visual processing system is doing something useful

Fig. 2.2 Slide-show RSVP
mode

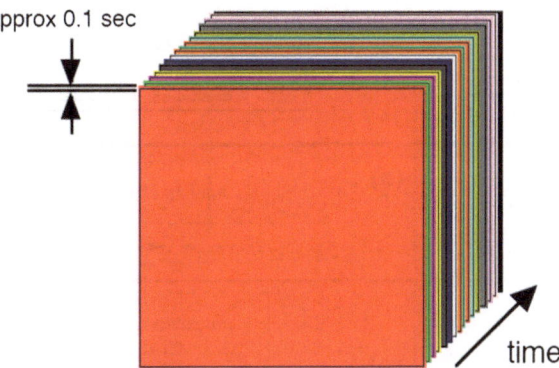

approx 0.1 sec

time

in a very short period of time, getting the "gist" of an image very rapidly. For easy reference and comparison with the other experiments discussed below we include the details of this first experiment in Table 2.1 as item (a).

The previous chapter pointed out that a person may not be seeking a known image but rather one that satisfies some criterion such as 'a cat' or 'an animal'. These more complex tasks may require the extended visibility of each image if the required target is to be reliably identified. Intraub (1981) and others (see Potter 1999, p. 17 top) determined that a longer viewing time (approx. 250 ms, equivalent to a presentation *pace* of 4/sec) is required, as summarised in part (b) of Table 2.1.

2.3 Pre-attentive Processing

Pre-attentive processing is illustrated by the experiment shown in Fig. 2.3 (Treisman 1985, 1991 and Treisman and Gormican 1988). A screen at first shows, for some seconds, a display (a) comprising a monochrome image. Then, for 150 ms, the display changes (b) to show a large number of blue circles and one red circle. Then (c) the display returns to its original state. It is found that even with such a short exposure, a subject will notice the presence of the red ('target') circle despite the presence of the many blue ('distractor') circles. If the presentation is repeated without a red circle, the subject is equally confident that no red circle was displayed.[1] The effect can reliably occur for exposure times as low as 50 ms.

The human visual behaviour illustrated in that experiment is called **pre-attentive processing**, because there is insufficient time in a period of (say) 150 ms to pay conscious attention to the display, and certainly insufficient time to conduct an exhaustive search: as we shall see in Chap. 4, a single eye movement event takes at least 200 ms to complete. The fact that pre-attentive processing is involved

---

[1] If you want to try this experiment go to www.csc.ncsu.edu/faculty/healey/PP/.

Table 2.1 Details of experiments

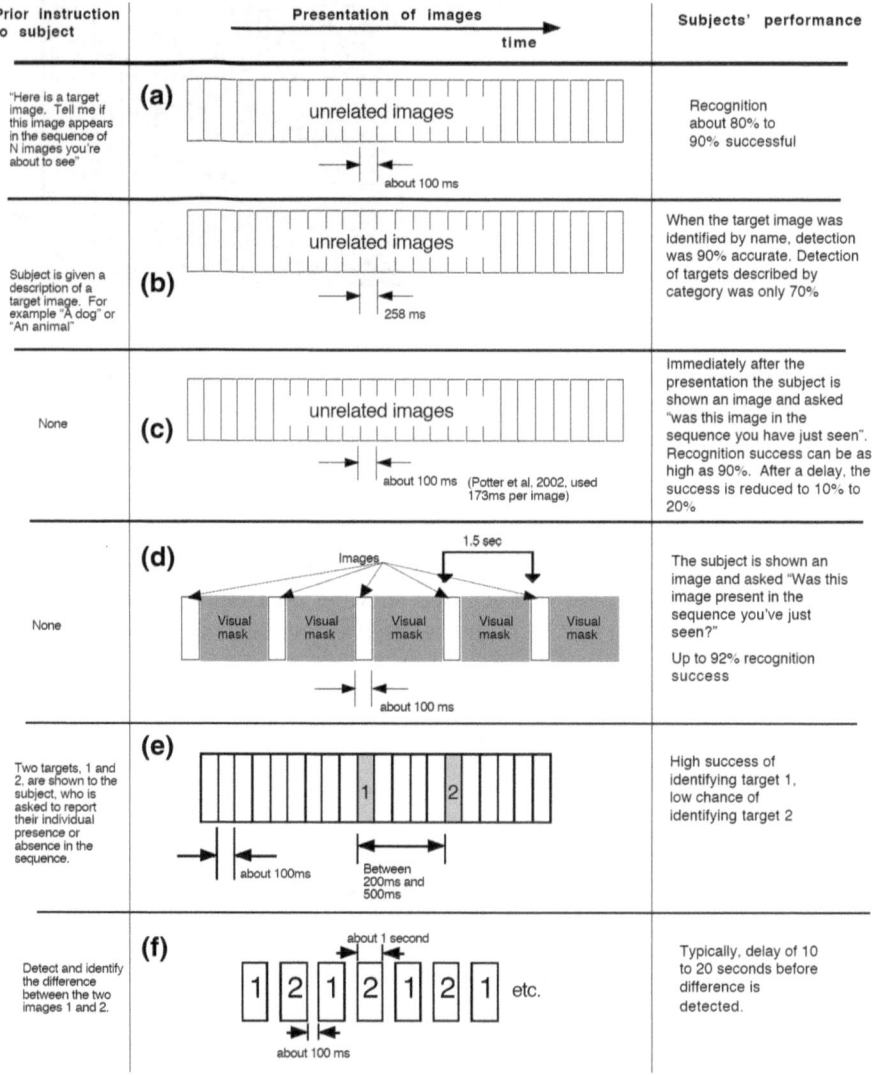

is demonstrated if the experiment is repeated with different numbers of distractors (represented diagrammatically in Fig. 2.4); success in recognising the presence or absence of a target is found to be essentially independent of the number of distractors. The phenomenon illustrated by the experiment is often referred to as **pop-out**. The target and distractors need not be circles, as Ware (2004) and Healey et al. (1996) point out.

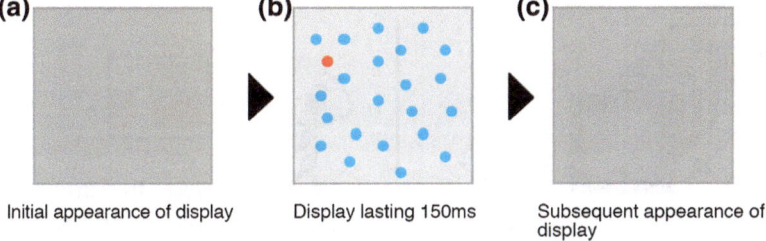

Fig. 2.3 Illustration of pre-attentive processing

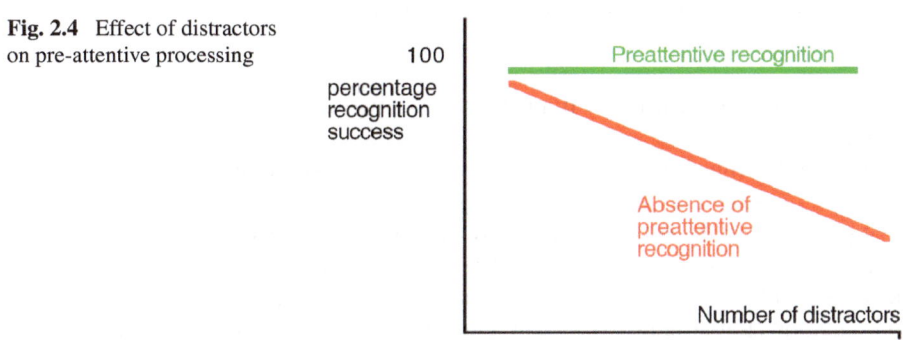

Fig. 2.4 Effect of distractors on pre-attentive processing

A second important feature of pre-attentive processing is that it involves no conscious cognitive effort (Potter 1999). Since something *useful* (target recognition) can occur within a *very short time* (e.g., 100 ms) and involve *no cognitive effort* it is no wonder that Colin Ware (2004) declared (page 149) that *An understanding of what is processed pre-attentively is probably the most important contribution that vision science can make to data visualization.*

The experiment just described employs, for good reason, an abstracted task, that of recognising a single simple target among a number of distractors. In contrast to abstract laboratory-based experiments, 'real world' examples of pre-attentive processing exist. For example, Kundel and Nodine (1975) briefly presented actual lung X-rays to trained radiologists for just 200 ms: surprisingly, 70 % of the target anomalies were still detected (performance rose to 97 % under unlimited viewing conditions).

2.4 Memory for Images

The remarkable results of the first two experiments certainly illustrate the potential of RSVP to help identify one image within a rapidly presented collection. What *cannot* be concluded from those experiments, however, is that 100 ms is also enough for the subject to *recall* those images at a later time: identify, yes, but

Fig. 2.5 Potter and Levy (1969) experiment: recall of rapidly seen image

remember, no, at least not with much success. Two experiments relate to the issue of memory.

Figure 2.5 illustrates an experiment carried out by Potter and her associates. A subject is shown a collection of images in Slide-show mode RSVP. Following that presentation the subject is then shown an image and asked to say whether that image was present in the collection. The success of identification could be as high as about 90 % if the question was put as little as 5 s after the termination of the presentation (Potter et al. 2002), but after that period the recall success fell rapidly to around 10 or 20 % (Potter and Levy 1969; Potter 1976). This experiment is summarised in Table 2.1c.

We can conclude from this experiment that interaction design using RSVP to support search for a given target within a collection should *not* assume that other images in the collection can later be recalled.

One might be tempted to conclude, from the experiment just described, that exposure to an image for 100 ms is insufficient for it to be remembered reliably and for some time. But one would be wrong! A remarkable experiment, illustrated in Fig. 2.6 and summarised in Table 2.1d, (Intraub 1980, 1984, 1999; Potter 1976) showed that if each brief (i.e., 100 ms) presentation of an image was followed for about 1.5 s by a 'visual mask'—for example a monochrome display—then the question "was this image in the collection you have just seen" was answered with

Fig. 2.6 Experiment with neutral display gaps between brief image presentations

up to 92 % success. This result suggests that whereas 100 ms would be sufficient for a user to *recognise* an image, it would be insufficient for its *consolidation* in short-term memory if followed immediately by another image.

2.5 Cognitive Blindness

There are periods when one is blinded briefly but without any conscious aware- ness of any interruption to the visual stream. Such "blindness" does not only result from the common activity of blinking: it can be brought about by very common actions such as shifting one's gaze from one location to another (saccadic blind- ness), or by being distracted by some more conspicuous change in the image (change blindness). Three experiments have established situations in which cogni- tive blindness can occur.

2.5.1 Attentional Blink

The phenomenon of attentional blink (Raymond et al. 1992; Kimron et al. 1999) refers to the fact that if, during a rapid presentation of a sequence of images, atten- tion is drawn to a particular image (it may, for example, be a target image) then attention is not available to identify or even detect a second target image for up to half a second after the successful report of the first target. The experiment and its result is described in Table 2.1e. The phenomenon has obvious implications for RSVP design, since the view of a familiar or very attractive image can impede search for a subsequent target image.[2]

2.5.2 Change Blindness

Change blindness is a very surprising phenomenon. In a typical experiment, a sub- ject is asked to view two different images that differ in some non-trivial respect, and to identify that difference. Thus, (see Table 2.1f) Image A might be shown for one second, then removed and replaced, after about 100 ms, by Image B which is itself shown for one second and similarly replaced by Image A, with the sequence repeating indefinitely. Experience shows that it may take a subject up to 10 or 20 s to notice the difference between Image A and Image B, a difference which, when noticed, is strikingly obvious and usually greeted by amusement. By accessing http://www.csc.ncsu.edu/faculty/healey/PP/ the reader can experience a number

---

[2] The reader can experience attentional blink by accessing the website: http://www.youtube. com/watch?v=MH6ZSfhdIuM.

Fig. 2.7 Example of mud splash distractors

of examples of this phenomenon which has been extensively reported and studied (see, for example, Rensink 2000; Rensink et al. 1997, 2000).

A variant of change blindness was illustrated by an experiment conducted by McConkie and Currie (1996). A photograph is shown (Fig. 2.7, left). It then changes (Fig. 2.7, right) but, at the same time, one or more distracting 'mud splashes' appear briefly (e.g., for a few 100 ms), as shown in Fig. 2.7, centre. Under these circumstances it can be extremely difficult to detect the change in the main image.[3]

Change blindness is not exclusive to computer-generated images. Coltheart (1999), for example, provides two different images of a building on opposite sides of a book's page, and much page turning is typically required to detect the difference. Another popular example arises from a study by Simons and Levin (1998). Here, the experimenter would stop passers-by on a college campus to ask them for directions, only to be suddenly replaced by a colleague when a visual obstacle (two men carrying a large board) passed between them. While the experimenter and his colleague were of different appearance, many people failed to notice the change.[4]

Why should change blindness be of interest in the context of RSVP? One example is provided by Example 1 of Chap. 1 where the collage mode of RSVP leads to the sudden appearance and disappearance of images in generally unpredictable locations. An example of its relevance to information visualization is provided by Nowell et al. (2001).

2.5.3 Saccadic Blindness

A particular form of change blindness was succinctly summarised by Chahine and Krekelberg (2009): "Humans move their eyes about three times each second. Those rapid eye movements—called saccades—help to increase our perceptual

---

[3] Actual examples for change blindness can be experienced through the website http://nivea.psycho.univ-paris5.fr/ECS/dottedline.gif.

[4] A similar, and amusing episode can be experienced at http://www.youtube.com/watch?v=ubNF9QNEQLA.

resolution by placing different parts of the world on the high-resolution fovea. As these eye movements are performed, the image is swept across the retina, yet we perceive a stable world with no apparent blurring or motion". This phenomenon is called saccadic blindness because it is associated with the very fast saccades that occur between two gaze locations.

The phenomenon was first described by Erdmann and Dodge in 1898, when it was noticed during unrelated experiments that an observer could never see the motion of their own eyes. The mind selectively blocks visual processing during eye movements in such a way that neither the motion of the eye (and subsequent motion blur of the image) nor the gap in visual perception is noticeable to the viewer. One subsequent experiment arranged for an image presented to a viewer to undergo change when the viewer's eyes underwent a saccade: many of the changes went undetected. To make matters worse, there appears to be evidence that saccadic blindness can start up to 50 ms *before* the onset of a saccade and last for 50 ms *beyond* its termination (Diamond et al. 2000).

2.6 Salience

In all of the fifteen examples discussed in Chap. 1 a human user is looking at a display. That single word 'looking' can mislead, because in most situations a user's gaze is constantly in motion: it dwells on a particular item for perhaps 200 ms, then quickly moves to another location where it may only dwell for 50 ms, and so on. Why? Because the makeup of the human eye only allows detail to be perceived within a visual angle of about two degrees: that's a thumb held at arm's length (try it). Consequently, to 'take in' a scene requires a number of such 'dwells'. The fascinating question, therefore, is "What decides the location of the next dwell"?

Basically, although many experiments have been carried out (see Witkowski and Randell 2007 for a review; also Findlay and Gilchrist 2003) the question is still not fully answered. Nevertheless, enough is known to provide a reasonably useful qualitative model.

The concept of *conspicuity* is relevant to the question. For example, a Master of Ceremonies' red dinner jacket among black tuxedos at a sombre state affair (Itti and Koch 2001) is conspicuous, and tends to attract attention automatically and without conscious effort on the part of a viewer. We can say that the red jacket 'pops out'.

The term *salience* denotes the relevance of a visual item to the task being undertaken. To extend the above example, if one is searching for the Master of Ceremonies at a formal dinner then the *colour* red associated with that person's jacket renders that visual item very *salient* to the search. That item may be additionally salient if the person is standing up among mainly seated diners. But of course there may be some seated ladies wearing red jackets, and others (waiters) without red jackets but who are standing up, so the decision where to look next must somehow assess the various degrees of salience of perhaps many items and prioritise one to which visual attention will next be paid.

A simple model is as follows (Findlay and Gilchrist 2003). A single item within an image (e.g., a red jacket) is characterised by two features. One is its *salience* to the task in hand (e.g., finding the Master of Ceremonies). In general, the salience of a given item may be due to a variety of features such as colour, intensity, shape, texture, orientation and movement. The other feature is an item's *proximity* to where the person's gaze is currently focussed: the proximity of an item to the current fixation increases that item's effective salience. Thus, for a given measure of item salience the item closest to the current gaze position will be the next point at which the user will probably look. Because a scene is composed of many items, each of which has its own salience for a given task, the concept of a salience map is often employed: that map is used to prioritise the various visual items in order to determine the location of the next fixation in a 'winner-takes-all' decision.

2.7 Implications for Design

From the experiments just described the interaction designer considering the use of Slide-show RSVP in an application should be aware that

- If the task is to recognise a well-defined specific target image within a collection, that collection can be presented at a rate as high as about 10 images per second without the recognition success rate falling below about 90 %;
- If the task is to detect a well-defined *type* of image (e.g., an animal) within a collection, then a presentation rate no higher than about 4 per second may be appropriate;
- More than about 6 s after seeing an RSVP presentation, the user should not be expected to remember whether a newly displayed image was in the collection;
- There is a danger, due to the phenomenon of attentional blink, that recognition of a target image might be compromised in a situation where two 'targets' or familiar images appear in close temporal proximity;
- Changes in a scene might not always be noticed;
- Excessive saccadic behaviour might lead to loss of recognition of visual changes.

References

Chahine, G., & Krekelberg, B. (2009). Cortical contributions to saccadic suppression *PlusOne*, *4*(9) (online).

Coltheart, V. (1999). *Fleeting memories cognition of brief visual stimuli*. Cambridge: MIT Press.

Diamond, M. R., Ross, J., & Morrone, M. C. (2000). Extraretinal control of saccadic suppression. *The Journal of Neuroscience, 20–9*, 3449–3455.

Erdmann, B., & Dodge, R. (1898). Psychologische Untersuchung über das Lesen auf experimenteller Grundlage. Niemeyer: Halle.

Findlay, J. M., & Gilchrist, I. D. (2003). *Active vision: The psychology of looking and seeing*. Oxford: Oxford University Press.

Healey, C. G., Booth, K. S., & Enns, J. T. (1996). High-speed visual estimation using pre-attentive processing. *ACM Transactions on Human Computer Interaction, 3–2,* 107–135.

Intraub, H. (1980). Presentation rate and the representation of briefly glimpsed pictures in memory. *Journal of Experimental Psychology: Human Learning and Memory, 6,* 1–12.

Intraub, H. (1981). Rapid conceptual identification of sequentially presented pictures. *Journal of Experimental Psychology: Human Perception and Performance, 7,* 604–610.

Intraub, H. (1984). Conceptual masking—the effects of subsequent visual events on memory for pictures. *Journal of Experimental Psychology: Learning, Memory, and Cognition, 10,* 115–125.

Intraub, H. (1999). Understanding and remembering briefly glimpsed pictures: Implications for visual scanning and memory. In V. Coltheart (Ed.), *Fleeting memories: Cognition of brief visual stimuli.* Cambridge: MIT Press.

Itti, L., & Koch, C. (2001). Computational modelling of visual attention. *Nature Reviews, Neuroscience, 2,* 194–203.

Kimron, L., Shapiro, K. L., & Luck, S. J. (1999). The attentional blink: A front-end mechanism for fleeting memories. In V. Coltheart (Ed.), *Fleeting memories: Cognition of brief visual stimuli.* Cambridge: MIT Press.

Kundel, H. L., & Nodine, C. F. (1975). Interpreting chest radiographs without visual search. *Radiology, 116,* 527–532.

McConkie, G. W., & Currie, C. B. (1996). Visual stability across saccades while viewing complex pictures. *Journal of Experimental Psychology: Human Perception and Performance, 22–3,* 563–581.

Nowell, L., Hetzler, E., & Tanasse, T. (2001). Change blindness in information visualization: A case study. *IEEE Proceedings of Information Visualization.*

Potter, M. C. (1975). Meaning in visual search. *Science, 187,* 965–966.

Potter, M. C. (1976). Short-term conceptual memory for pictures. *Journal of Experimental Psychology-Human Learning and Memory, 2,* 509–522.

Potter, M. C. (1999). Understanding sentences and scenes: The role of conceptual short-term memory. In V. Coltheart (Ed.), *Fleeting memories: Cognition of brief visual stimuli.* Cambridge: MIT Press.

Potter, M. C., & Levy, E. I. (1969). Recognition memory for a rapid sequence of pictures. *Journal of Experimental Psychology, 81,* 10–15.

Potter, M. C., Staub, A., Rado, J., & O'Connor, D. H. (2002). Recognition memory for briefly presented pictures: The time course of rapid forgetting. *Journal of Experimental Psychology-Human Perception and Performance, 28,* 1163–1175.

Raymond, J. E., Shapiro, K. L., & Arnell, K. M. (1992). Temporary suppression of visual processing in an rsvp task—an attentional blink. *Journal of Experimental Psychology: Human Perception and Performance, 18,* 849–860.

Rensink, R. A. (2000). The dynamic representation of scenes. *Visual Cognition, 7,* 17–42.

Rensink, R. A., O'Regan, J. K., & Clark, J. J. (1997). To see or not to see: the need for attention to perceive changes in scenes. *Psychological Science, 8,* 368–373.

Rensink, R. A., O'Regan, J. K., & Clark, J. J. (2000). On the failure to detect changes in scenes across brief interruptions. *Visual Cognition, 7*(1–3), 127–145.

Simons, D. J., & Levin, D. T. (1998). Failure to detect changes to people during a real-world interaction. *Psychonomic Bulletin and Review, 5–4,* 644–649.

Treisman, A. (1985). Pre-attentive processing in vision. *Computer Vision, Graphics and Image Processing, 31,* 156–177.

Treisman, A. (1991). Search similarity and integration of features between and within dimensions. *Journal of Experimental Psychology: Human Perception and Performance, 17*(3), 652–676.

Treisman, A., & Gormican, S. (1988). Feature analysis in early vision: evidence from search asymmetries. *Psychological Review, 95–1,* 15–48.

Ware, C. (2004). *Information visualization: perception for design.* Amsterdam: Morgan Kaufmann.

Witkowski, M., & Randell, D. A. (2007). A model of modes of attention and inattention for artificial perception. *Bioinspiration and Biomimetics, 2,* S94–S115.

Chapter 3
RSVP Modes and Their Properties

Abstract This chapter looks in detail at the range of RSVP modes used today. They are broadly categorised into (1) static modes, where images appear and disappear without moving, (2) moving modes, where images appear sequentially, move about the display and then disappear, and (3) multiple entry/exit modes, in which images appear and disappear in many locations or move along several paths. We consider how design parameters such as presentation rate, speed of movement across the display, image size and overlap, as well as the use of manual control can influence the effectiveness and attractiveness of an RSVP mode. These discussions are richly illustrated with examples from experiments. To assist with the development process we introduce a graphical design notation for RSVP modes.

Keywords RSVP design modes • Design notation • Static mode RSVP • Moving mode RSVP • Multiple entry/exit RSVP

The purpose of Chap. 1 was to illustrate the wide variety of applications in which the potential offered by RSVP might be exploited. The present chapter now examines various visual and dynamic forms that RSVP can take, and the properties that characterise them in terms of task support. Some of the RSVP 'modes' discussed in this chapter will be familiar from Chap. 1.

3.1 Descriptions

It is frequently necessary to *describe* a particular mode of RSVP with some precision, especially when an interaction designer has to specify the nature of an application to the person responsible for its coding. For something that is essentially visual, words are often—and unsurprisingly—inappropriate. Even a picture, however, can have limitations. Consider, for example, the screen shot of Fig. 3.1 containing a number of images. Many questions occur that cannot be resolved by this description: "where do images appear?"; "at what rate do they appear?"; "do they move?"

R. Spence and M. Witkowski, *Rapid Serial Visual Presentation*, SpringerBriefs
in Computer Science, DOI: 10.1007/978-1-4471-5085-5_3, © The Author(s) 2013

Fig. 3.1 Appearance of
volcano mode RSVP

"If so, in what direction(s), how fast, and smoothly or otherwise?"; "If the images
move, do they 'freeze' momentarily at some location?"

To solve the problem of description, and be able to answer such questions, we
use a notation specifically developed for this purpose. It is illustrated in the context of the Volcano mode RSVP introduced in Chap. 1 (Fig. 1.9) and whose screen
appearance has just been seen in Fig. 3.1. The corresponding notational description shown in Fig. 3.2 uses image outlines as its basis and shows (a) by a solid circle where all images appear; (b) by empty circles where they disappear from view;
(c) by a line with arrows to indicate the trajectory of image movement; (d) by a
'right-angle' symbol ('¬') on each image outline to indicate continuous rather then
discrete movement; and (e) by a cross to indicate where an image is static ('captured') for a short period of time. If the empty circles appear outside the display
boundary the images *gradually* disappear from view, otherwise they disappear
abruptly: the same rule applies for image appearance. Only three items of quantitative information are needed to complete such a description: the rate at which
images appear (we refer to this as 'pace'), the speed of image movement and the
period for which an image is captured.

Fig. 3.2 Notational
representation of volcano
mode

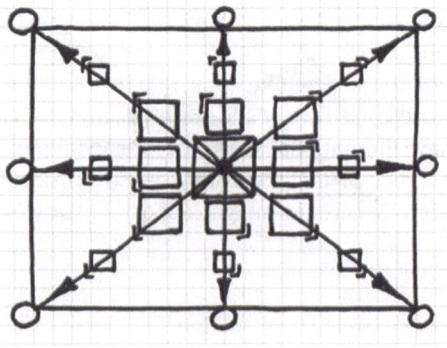

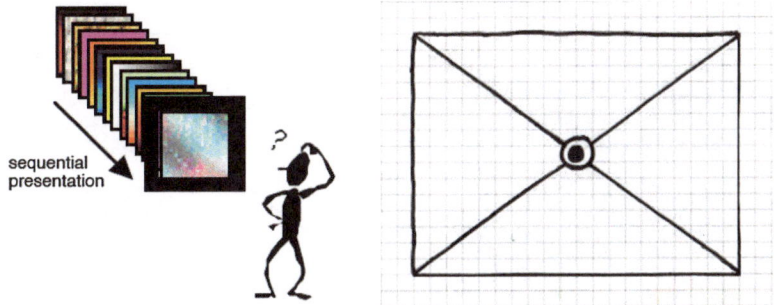

Fig. 3.3 Slide-show mode—concept and notational description

A simple example of the notation's value is provided by Fig. 3.3 which shows the earlier illustration of Slide-show RSVP and its notational representation. The latter may appear somewhat ascetic but is unambiguous and compact.

3.2 An Abundance of RSVP Modes

As demonstrated in Chap. 1, many modes of RSVP are possible, so it is natural to ask whether any useful taxonomy exists that simplifies consideration of the properties that the various modes exhibit. It is found to be convenient to separate RSVP modes into three classes: Static, Moving and Multiple entry/exit.

3.2.1 Static Modes

In the static group of RSVP modes, images have *identical entry and exit locations*, and images are replaced at a constant rate. Slide-show mode RSVP (Fig. 3.4) is an example where each image occupies the entire display area. As already mentioned, an advantage is that a target image is likely to be identified even if only visible for 100 ms. However, a potential disadvantage of this mode is that there is no opportunity after that time to confirm recognition. Also, an eye-blink—typically of 300–400 ms duration—can obviously lead to one or more images not being seen.

To allow more time for confirmation of a target's recognition, but *without extending the overall image collection presentation duration*, the Slide-show mode can be extended to multiple concurrent images as illustrated by the 'Mixed' mode shown in Fig. 3.5. In this mode, each image is displayed for four times the previous duration (e.g., 400 ms) but at a quarter of the previous image size. The reason for considering this Mixed mode lies in the notion of resource, in this case the combination of display area and total image collection presentation time. In typical situations the display area is specified and the interaction designer would be

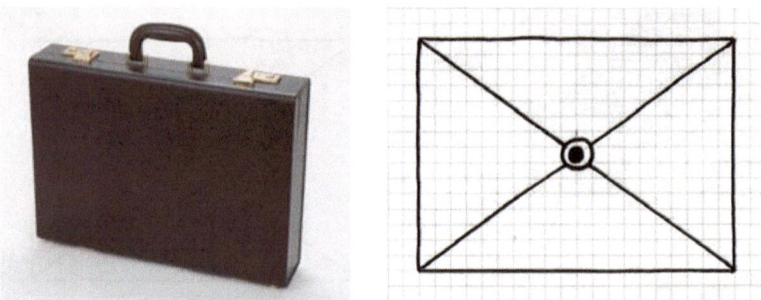

Fig. 3.4   The static slide-show RSVP mode: (*left*) appearance, (*right*) notation

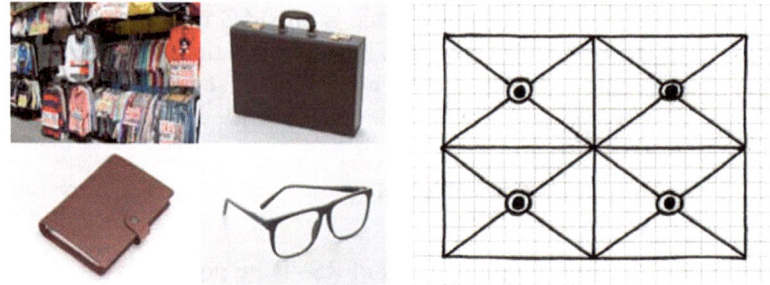

Fig. 3.5   The mixed (2 × 2) RSVP mode: (*left*) appearance, (*right*) notation

reluctant to occupy the user for longer than is necessary to complete a task. Thus, a fixed resource requires a 'trade-off' between image size and individual image presentation time and is a reasonable basis on which to compare static RSVP modes.

We examine the properties of static modes below in Sect. 3.3.1 after considering two other types of RSVP, namely moving and multiple entry/exit.

3.2.2 Moving Modes

A user can be given an opportunity to confirm their decision about any particular image by retaining it on the display but moving it to make room for subsequent images in a collection. Three of many possible examples of this mode are shown in Figs. 3.6, 3.7 and 3.8, in each case showing both the appearance (left) and notational description (right). In Diagonal mode (Fig. 3.6) images appear top left, then move continuously towards the bottom right, where they are 'captured' (i.e., remain fixed in that location, as indicated by the cross) for (say) 100 ms and then disappear, as indicated by the open circle. The image route is linear and

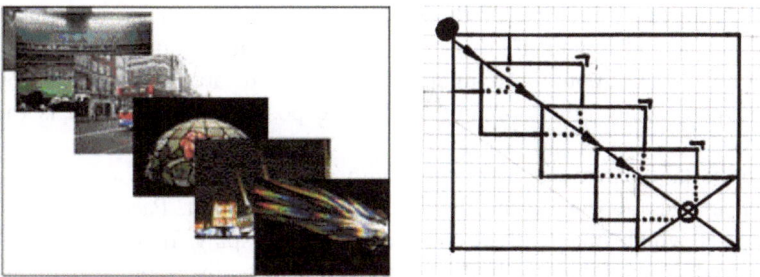

Fig. 3.6 Diagonal mode RSVP: (*left*) appearance, (*right*) notation

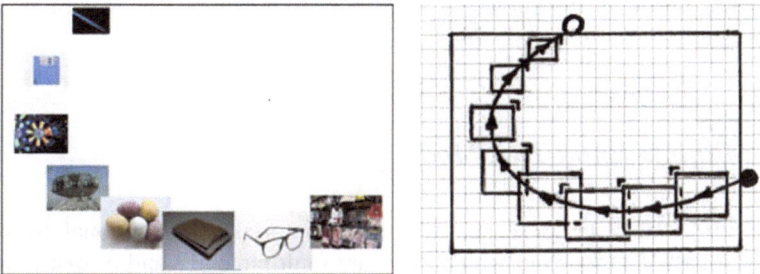

Fig. 3.7 Stream mode: (*left*) appearance, (*right*) notation

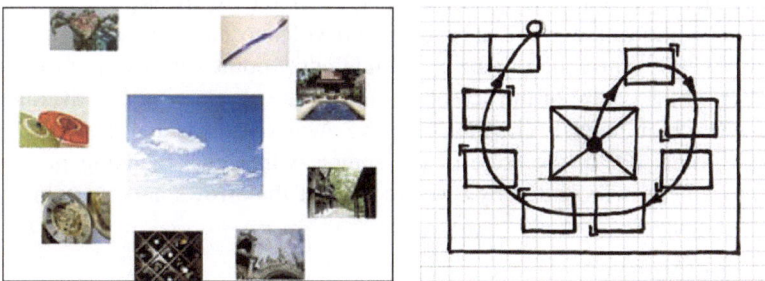

Fig. 3.8 Ring mode: (*left*) appearance, (*right*) notation

of constant speed, as indicated by the straight arrow and equal spacing between image outlines.

Figure 3.7 provides an example of the flexibility available to the interaction designer who may, for artistic or other reasons, wish to use a moving mode of RSVP. It shows the Stream mode in which the image stream varies in both size and speed, indicated respectively by the changing size and spacing of the image outlines. Here there is no capture frame; images appear smoothly at the entry point to the right and leave the display area continuously at the top.

Another form of moving mode is shown in Fig. 3.8. It shows Ring mode; images appear and remain for (say) 100 ms in a central capture frame and then move continuously in an approximately spiral path, disappearing off the top edge of the screen. Many other moving modes are possible: imagination is the only limit.

An important parameter associated with any mode is that of *pace*—the rate at which images appear on the display at the entry point. Pace is distinct from the *speed* at which images then move across the display. If, in Diagonal mode, image speed is insufficiently high, the images will be overlapped as in Fig. 3.6: conversely, with either a low pace and/or high image speed, images will be separated as they move within the display (as in, for example, Figs. 3.7 and 3.8). With overlap, the illusion of a third dimension, 3D depth, can be simulated, as shown in Fig. 3.6, and may be attractive in certain applications in view of its visual seductiveness.

3.2.3 Multiple Entry/Exit Modes

Figures 3.9, 3.10, 3.11, 3.12 show, in both pictorial and notational form, four examples of RSVP modes in which there are multiple entry and/or exit points for images, possibly connected by multiple movement paths. Again, as with moving modes, the principal anticipated advantage is the opportunity to confirm recognition of a target image. These designs tend to have a large number of images simultaneously on the screen, with movement and overlap in many cases.

Figure 3.9 shows the previously discussed Volcano mode (Corsato et al. 2008; Porta 2006, 2009), in which there is a single entry point with capture (notated by '●' and ×) at the centre of the screen at a moderate image size (e.g. average diagonal of 140 screen pixels). The images move away outwards continuously towards eight exit points (O) at the edge of the screen, while also reducing in size. The image movement is indicated by the direction of the arrows and changing size of the outline images.

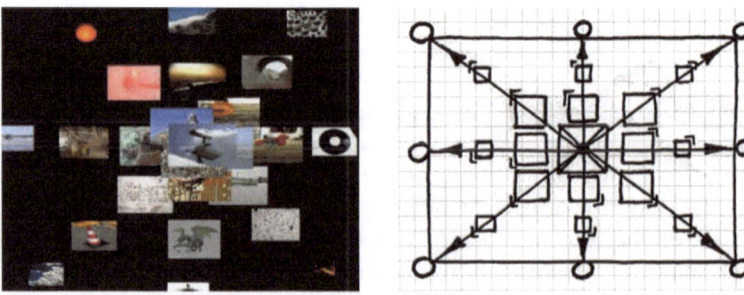

Fig. 3.9 Volcano mode: (*left*) appearance, (*right*) notation

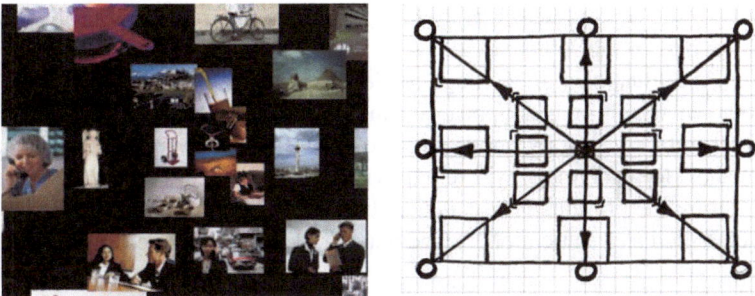

Fig. 3.10   Floating mode: (*left*) appearance, (*right*) notation

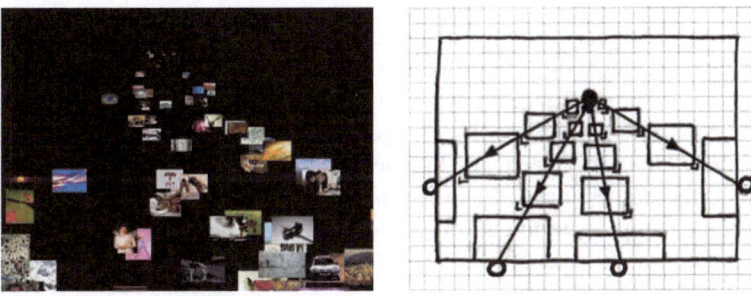

Fig. 3.11   Shot mode: (*left*) appearance, (*right*) notation

Fig. 3.12   Collage mode: (*left*) appearance, (*right*) notation

A similar mode is the Floating mode (Fig. 3.10), in which the images appear in miniature at the centre of the screen, moving outwards as in Volcano mode, but getting larger towards the exit points at the edge of the screen, thereby creating the 'motorway billboard' effect (see the shopping Example 3 in Chap. 1).

Figure 3.11 illustrates Shot mode (Corsato et al. 2008; Porta 2006, 2009), emulating images 'fired' from a single entry point, subsequently moving along many paths and increasing in size prior to leaving continuously at exit points around the lower part of the screen area. Unlike Floating mode, the radial paths are restricted to a smaller part of the screen area.

Figure 3.12 illustrates the Collage mode (Wittenburg et al. 1998, 2000; Corsato et al. 2008) in which images appear at random locations on the screen and remain there either until covered by another image or until the sequence ends. There is no guarantee how long an image will remain visible, or the extent to which another may cover it during the presentation sequence. It is true that Collage mode could equally well be classified as a static mode, since none of the images move on the screen.

3.3 Properties of Modes

An interaction designer contemplating the use of RSVP in an application will naturally want to know about the properties of different modes. "How easy is it for a user to find a known image?"; "Does it matter if some images overlap?"; "How useful is 'capture', when a moving image is temporarily halted?"; "Do users like this mode?"; "Do users experience fatigue when using this mode?"

Some of these questions refer to our *objective* knowledge about a mode's properties obtained from experiment, properties such as the success in recognising the presence of an image in a collection and the ability to recall images that have been presented in a rapid sequence.

Other questions refer to *subjective* opinions elicited from users, such as perceived task success and fatigue. One important subjective property, especially when designing for commercial applications, is visual appeal: this may well be a major determinant of design choice but is not easy to quantify; it is difficult to elicit and classify and to a large extent relies on the creativity of the designer. Yet other questions refer to issues for which only opinions are available, such as the effect of capture frames. It turns out that both objective and subjective mode properties are most conveniently discussed together for each of the three classes of RSVP mode.

3.3.1 Static Modes

As Chap. 2 has pointed out, Slide-show RSVP has been extensively studied by psychologists (for a comprehensive review see Coltheart 1999) and the results of experiments are well documented. Briefly, it is known that recognition of a target image can occur with about 90 % success even when the pace of image presentation is as high as 10/s (see Table 2.1a in Chap. 2), though a slower pace may be needed if targets are described by category (see Table 2.1b). It has also been established that memory for presented images may only be satisfactory within a few seconds after the rapid presentation of an image collection (Table 2.1c). A significant drawback to Slide-show mode, of course, is that there is no opportunity after the disappearance of an image to confirm recognition, though manual control over the rate and direction of presentation can ameliorate this potential drawback.

An investigation (Spence et al. 2004) of the extension of Slide-show mode to allow smaller images to be presented for a proportionately longer time is represented in Fig. 3.13 and includes, for comparison, the Tile mode: this was included, not as an RSVP mode, but as a natural extension of the trade-off between image size and presentation time while maintaining a constant value of the area and time resources. The experiment was intentionally conducted with very small presentation times in order to identify limits to advisable operation: the image presentation times for Slide-show mode varied from 15 to 90 ms.

The experimental results for the three static modes of Fig. 3.13 are shown in Fig. 3.14 for three different sizes of display, roughly corresponding to a regular monitor ('large'), a PDA/tablet ('medium') and a small mobile ('small'). The provision of three separate abscissa scales is intentional. Below the total image collection presentation time (T) scale is shown the presentation time per image (T_A) for which each individual image is visible. Intuitively (but see below) it may be expected that, the greater T_A, the greater would be the recognition success. The lower scale is a parameter T_i which is simply the total presentation time divided by the number of images.

From the evidence presented in Fig. 3.14 it would appear that, at least for the three static modes investigated, all begin to be associated with significant image recognition error as the value of T_i falls below about 100 ms, leading to the guideline that total presentation time for a collection of N images should be no less than N/10 s and should preferably comfortably exceed this value.

Intuitively, one might conjecture that the likelihood of recognition success is increased the longer that an image is visible. Thus, it was expected that, unless the images would then be too small, the mixed mode would lead to greater identification success. This suggestion appears to be the case for T_i values less than 100 ms.

It is also confirmed for the Slide-show and Mixed modes by the result shown in Fig. 3.15 plotted for the large display. However, also plotted in Fig. 3.15 is the result for Tile mode, which apparently contradicts the general conjecture and thereby provides a cautionary tale for the interaction designer. The most likely explanation for the apparently confusing evidence is—as we discuss soon in Chap. 5—that

Fig. 3.13   Space—time trade-off in static mode (slide-show, mixed and tile)

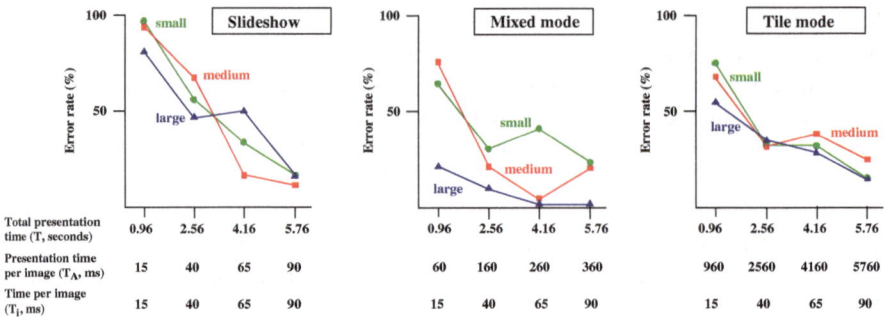

Fig. 3.14 Experimental results for static modes (after Spence et al. 2004)

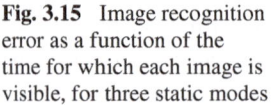

Fig. 3.15 Image recognition error as a function of the time for which each image is visible, for three static modes

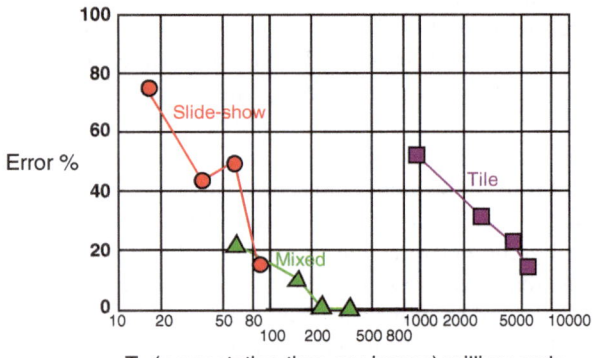

Slide-show and mixed modes, at the values of T_A investigated, result mainly in pre-attentive processing, whereas in tile mode more attentive search is involved .

In the same experiment users expressed an overwhelming preference for the Mixed mode. Though no application has yet been reported for the 2×2 Mixed mode, intellectual curiosity remains concerning the possible advantages of a 3×3 presentation permitting even longer inspection times.

When considering the potential of Slide-show mode the interaction designer must be aware of wider implications in addition to the abstract laboratory investigations reported above, and also think about situations where Slide-show mode is simply unsuitable. An illustrative example is provided by the Flipper application (Sun and Guimbretiere 2005) where (see example 5 in Chap. 1), in the examination of a text, it may well be beneficial for Slide-show RSVP mode at high page-turning rates to be complemented automatically by scrolling at lower rates. Similarly, the professional engaged in video editing may well wish to search in a Slide-show mode at very high presentation rates but in an environment where context is visible and where manual control can provide both fast exploration and slower detailed examination.

3.3.2 Moving Modes

What can moving modes offer that static modes cannot? An obvious potential advantage is that time is then available for a user to confirm recognition of a target image, but this assertion needs to be investigated by experiment. Even if the assertion is true, are there concomitant disadvantages for moving modes?

An investigation by Cooper et al. (2006) throws some light on this issue and, at the same time, suggests the profound influence on recognition behaviour of eye-gaze, a topic discussed at length in the following chapter. In the investigation the six presentation modes shown in Fig. 3.16 were used to present collections of 48 images, with total presentation times ranging from 3.36 to 6.24 s. Again, presentation times were intentionally chosen to investigate a suspected boundary between acceptable and unacceptable performance, with image visibility in Slide-show mode varying from 70 to 130 ms. Tasks undertaken by subjects ranged from simple target recognition to the recognition of image themes (e.g., 'ships') and categories (e.g., animals).

It was found that, over all modes, presentation times and tasks, recognition accuracy averaged 84 % across all static modes (Slide-show, Mixed, Tile) and 62 % across all moving modes (Diagonal, Ring, Stream). It was also found that target recognition accuracy for static modes was relatively little influenced by the nature of the task, in contrast to the outcome for moving modes. Seventy-five percent of subjects preferred the static modes.

These results need to be interpreted with care. First, the individual image presentation times (70–130 ms) were intentionally chosen to examine any boundary between acceptable and unacceptable performance. Second, it could validly be claimed that the selection of modes for investigation, especially the moving ones, was somewhat arbitrary. The fact that separation into 'static' and 'moving' modes

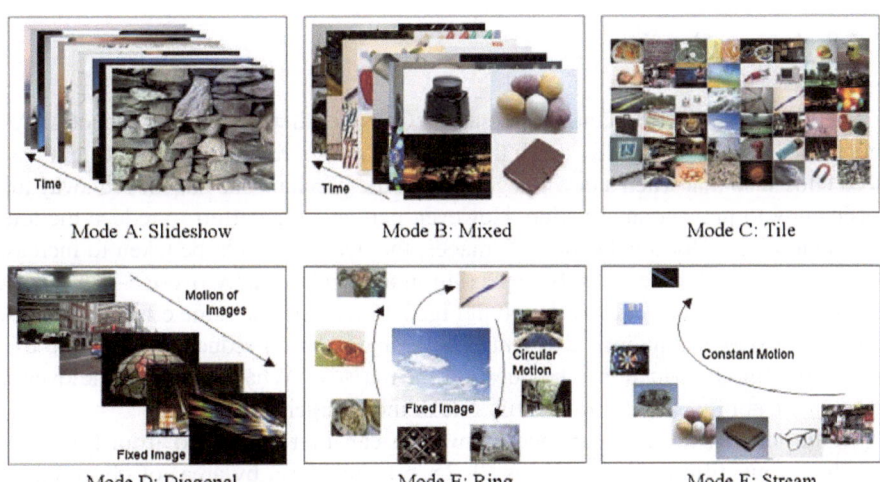

Mode A: Slideshow   Mode B: Mixed   Mode C: Tile

Mode D: Diagonal   Mode E: Ring   Mode F: Stream

Fig. 3.16 RSVP modes investigated by Cooper et al. (2006)

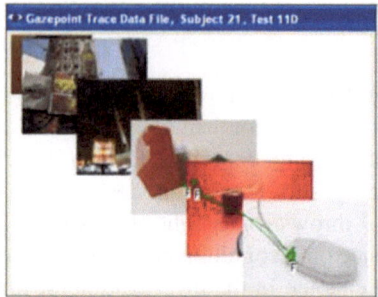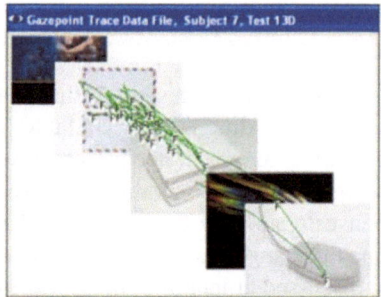

Fig. 3.17 Likes (*left*) and dislikes (*right*) relating to gaze travel (reprinted from Cooper et al. 2006)

is an oversimplification is demonstrated by records of the eye-gaze behaviour of subjects carrying out tasks using the different presentation modes, as we discuss immediately below and later in Chaps. 4 and 5 in more detail.

Take the Diagonal mode, for example. Figure 3.17 shows the eye-gaze records of two different users. For one user (Fig. 3.17, right) gaze, shown by the green line superimposed over the image, is continuously moving up and down the trajectory followed by the images. For another user (Fig. 3.17, left), gaze is largely concentrated at the point at which the disappearing images are captured for a short period, effectively viewing the image collection in Slide-show mode!

Does this difference in visual technique matter? Apparently yes: it is recorded that the viewer associated with Fig. 3.17 (left) liked the Diagonal mode, whereas the viewer associated with Fig. 3.17 (right) disliked it.

3.3.3 Image Overlap

Image overlap is an important consideration. Consider first the Diagonal mode RSVP represented in Fig. 3.18 (left). Both the pace and the speed of image movement have been chosen so that when a new image appears the previous one has moved away just sufficiently to prevent any overlap. However, perhaps in an effort to reduce the total presentation time for a collection of images, the decision might be taken to increase the pace, in which case a newly appearing image will partially overlap—and hence obscure—the previous one, as shown in Fig. 3.18 (right). The same effect could arise if the pace were maintained constant but the image speed reduced to give the user more time for confirmation. The question then arises, "What effect will the overlap have on the ability of a user to identify one or more target images?"

Some idea of the consequence of overlap can first be gained from Fig. 3.19, in which four different percentages (0, 25, 50 and 75 %, by area) of obscurement are shown. For the simple image chosen for illustration it is seen that a considerable degree of obscurement might be tolerated. On the other hand, a more detailed

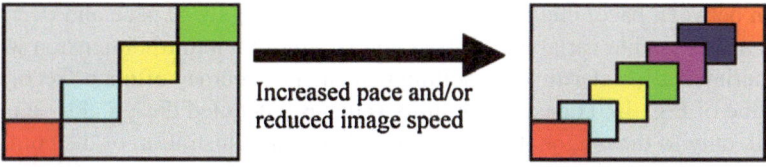

Fig. 3.18   Effects of pace and speed on overlap in diagonal mode RSVP

Fig. 3.19   Visual effect of different levels of obscurement (reprinted from Brinded et al. 2011)

Fig. 3.20   Recognition accuracy as a function of pace and overlap (reprinted from Brinded et al. 2011)

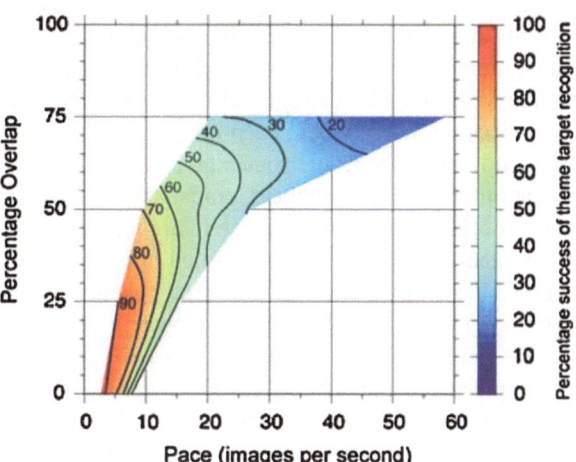

image, especially in the context of similar images in a sequence, might not be so tolerant of obscurement.

A study (Brinded et al. 2011) of the effect of overlap on image recognition success in Diagonal mode RSVP led to the representation of Fig. 3.20. As the discussion above, and the illustration of Fig. 3.18, has shown, there is a very simple relation between pace, image speed and overlap: in Fig. 3.20 pace and overlap are treated as designable variables while contours represent a third dimension which is the experimentally determined resulting percentage accuracy of image recognition. The value of Fig. 3.20 does not lie in its value as a detailed design aid—it applies, after all, only to the diagonal mode—but rather as an illustration of the complexity of the effect of overlap, a complexity leading to interesting trade-offs. For example, if 70 % theme recognition accuracy is acceptable there is a choice between a pace of 12 images per second with 30 % overlap and a pace of 9.5 images per second with an overlap of 50 %.

3.3.4 Multiple Entry/Exit Modes

The interaction designer might understandably ask what potential advantages are offered by more complex RSVP modes such as Volcano, Floating, Shot and Collage, and at the same time wonder about any design constraints they might impose. A useful response to such questions can be gained from studies of these four representative modes by Corsato et al. (2008) and Porta (2006, 2009).

The screen appearance and notational description of these four modes, already shown in Figs. 3.9, 3.10, 3.11 and 3.12 are repeated as the first two columns of Fig. 3.21. In the third column of this figure the nature of gaze movement for each of the four modes is illustrated by gaze traces (shown in yellow, here). Each gaze trace shows, for one viewing session lasting some 224 s, the movement of gaze activity over the display area. In each case the gaze track was recorded from the same individual while viewing each of the different modes.

For Volcano and Floating modes we see a concentration of gaze at the centre (possibly because that is where each image is 'captured' for a short time) but with some gaze movement along the routes taken by images, presumably to confirm the identification of a desired image. Compared with Volcano and Floating modes the gaze track for Shot mode differs in one major respect: gaze is concentrated in a larger area associated with the locations at which images can first be identified. Unsurprisingly, with Collage mode, gaze travel is pretty evenly spread over the display, reflecting the arbitrary locations at which images appear. Column three of Table 3.1 shows the total gaze travel (measured in screen pixels) associated with the four modes.

A natural question to ask is "Does the extent of gaze travel matter?" It appears that that is the case. Given the task of identifying thematic images (e.g. "cat" or "ship") from a selection of 2000, Table 3.1 shows, for each of the four modes, (1) the success in the recognition task (out of a possible 40 target images), (2) total gaze travel, and (3) a measure of fatigue (a scale from 1, no fatigue, to 5, extreme fatigue). The very strong suggestion emerging from these and all other examples of gaze behaviour examined in this chapter is that the more extensive the gaze travel, the lower the recognition success and the greater the fatigue felt by a user. This suggestion and other observations arising from the careful analysis of gaze data are examined in detail in the following chapter.

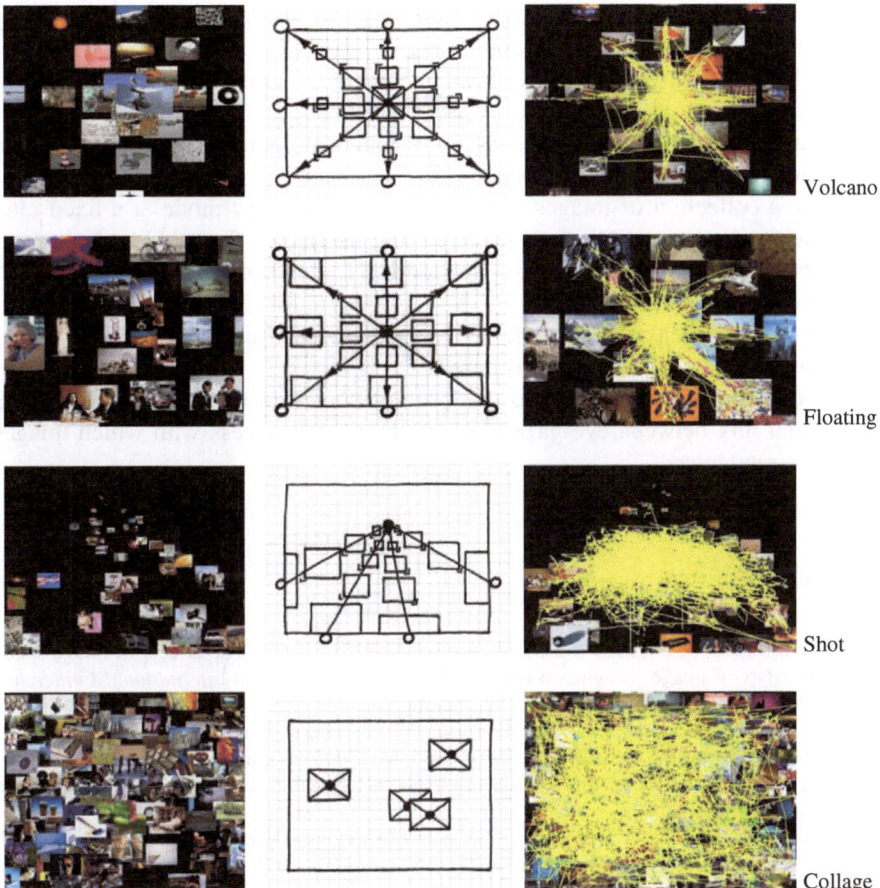

Fig. 3.21 Notation and overall gaze travel for various multiple entry/exit modes

Table 3.1 Multiple entry/
exit modes: recognition, gaze
travel and fatigue measures
(data from Corsato et al.
2008)

Mode	Recognition success (max 40)	Total gaze travel (pixels)	Degree of fatigue (1–5)
Volcano	28.00	14,810	2.71
Floating	29.51	13,462	2.52
Shot	19.29	25,783	3.13
Collage	19.35	36,349	3.90

3.4 Design Considerations

Although Chap. 6 will specifically address the design of an RSVP application, it
is nevertheless useful to the interaction designer to summarise the issues that have
been developed in this chapter:

- Examine the task(s) associated with the application to see when RSVP may offer benefits and when it might not.
- Establish whether exploration or search or a combination of the two tasks will be carried out.
- The availability of context presentation and manual control of rate and direction may influence the choice of mode.
- Where a collection of images is presented in Slide-show mode at a fixed rate, and single target recognition is involved, the parameter T_i should be kept comfortably above 100 ms. Thus, a collection of N images should not be presented in a total time less than N/10 s.
- As the task increases in complexity (e.g., theme and category targets) slower presentation rates should be considered.
- In anticipation of more detailed guidance presented in Chap. 5, be aware of a potential link between eye-gaze movement and the success with which images may be recognised.

References

Brinded, T., Mardell, J., Witkowski, M., & Spence, R. (2011, July). The effects of image speed and overlap on image recognition (pp. 3–11). *Proceedings of 15th International Conference on Information Visualization (IV2011)*, London.

Coltheart, V. (1999). *Fleeting memories: Cognition of brief visual stimuli*. Cambridge: MIT Press.

Cooper, K., de Bruijn, O., Spence, R., & Witkowski, M. (2006). A comparison of static and moving presentation modes for image collections (pp. 381–388). *Proceedings of Advanced Visual Interfaces (AVI-2006)*.

Corsato, S., Mosconi, M., & Porta, M. (2008, May). An eye tracking approach to image search activities using RSVP display techniques, (pp. 416–420). *ACM Proceedings Workshop on Advanced Visual Interfaces*, Naples.

Porta, M. (2006). Browsing large collections of images through unconventional visualization techniques, ACM, (pp. 440–444). *Proceedings AVI 2006*.

Porta, M. (2009). New visualization modes for effective image presentation. *International Journal of Image Graphics, 9–1*, 27–49.

Spence, B., Witkowski, M., Fawcett, C., Craft, B., & de Bruijn, O. (2004). Image presentation in space and time: errors, preferences and eye-gaze activity, (pp. 141–149). *ACM Proceedings of Workshop on Advanced Visual Interfaces (AVI-04)*.

Sun, L., & Guimbretiere, F. (2005). Flipper: A new method for digital document navigation, (pp. 2001–2004). *ACM Proceedings CHI'05 (Extended Abstracts)*.

Wittenburg, K., Ali-Ahmad, W., LaLiberte, D., & Lanning, T. (1998). Rapid-fire image previews for information navigation, (pp. 76–82). *ACM, Proceedings of Conference on AVI*.

Wittenburg, K., Chiyoda, C., Heinrichs, M., & Lanning, T. (2000, January 26–28). Browsing through rapid-fire imaging: Requirements and industry initiatives (pp. 48–56). *Proceedings of Electronic Imaging '2000: Internet Imaging*, San Jose, CA, USA.

Chapter 4
Eye-Gaze

Abstract This chapter considers the principles of eye movements and how eye-gaze recording and analysis techniques can be used to study and understand the consequences of design decisions. First we look at the different ways in which gaze naturally reacts to different RSVP-like presentations and describe four distinct gaze behaviours sufficient to characterise the significant aspects of eye movements observed during RSVP sequences: (1) visual search, (2) steady gaze, (3) nystagmus, and (4) visual pursuit. The chapter concludes with a discussion of ways of representing and presenting eye-gaze information to best effect.

Keywords Eye-gaze recording • Saccades • Fixations • Visual search • Nystagmus • Visual pursuit • Fovea • Representing eye-gaze data

In the course of the last chapter we saw that users often tend to prefer 'static' RSVPs, where each image exists in only one location, rather than 'moving' RSVPs in which a given image can move around the display. We also saw that the search for a target image tended to be more successful with static presentation modes than moving ones. One obvious question to ask is "Why?"—and, equally, "what can an understanding of eye-gaze contribute to RSVP design?"

4.1 Why is Eye-Gaze Important in RSVP Design?

To seek an answer, Cooper et al. (2006) and others have recorded the eye-gaze of subjects given the task of identifying the presence or absence of a target image within a displayed collection of images. How gaze is detected and recorded is discussed later in this chapter when we know what it is we need to discuss: suffice it to say for now that eye-gaze is typically 'on the move', controlled by muscles around the eyeball, often in very characteristic ways. It is well known that a person's gaze adapts not only to the visual environment, but also to the activities and tasks they are performing (e.g. Yarbus 1967, for a classic demonstration).

R. Spence and M. Witkowski, *Rapid Serial Visual Presentation*, SpringerBriefs   47
in Computer Science, DOI: 10.1007/978-1-4471-5085-5_4, © The Author(s) 2013

In normal image viewing conditions, gaze will typically be directed to one location on a display screen for about 300 ms (this is termed a *fixation*) and will then move to another location extremely rapidly, for example in about 20 ms (that movement is termed a *saccade*). Various RSVP modes, however, give rise to a wide range of different gaze behaviours, which in turn can have an impact on how the mode is viewed, how long the images must remain visible, how big the images must be, how effective the mode is, and the degree to which the user will enjoy using it.

Users are generally unaware of their own eye movements, but an understanding of gaze behaviour can give the interaction designer an indication of what will be noticed by the user and what might be missed, as well as some insight into the degree to which a user might prefer using a particular mode of presentation.

4.2 Gaze Strategies

In this section we consider four distinct eye-gaze strategies that people use when looking at different RSVP modes: visual search, steady gaze, nystagmus and visual pursuit. When viewing any particular RSVP mode users will immediately adopt one of these gaze strategies, or a combination of them, in characteristic ways for the duration of the presentation.[1] Later in Chap. 5 we examine ten different RSVP modes and describe and classify each in terms of these four types of gaze behaviour.

4.2.1 Visual Search

An example of gaze behaviour typical of a person viewing a static picture is provided in Fig. 4.1 which shows a sequence of fixations (coloured cyan and denoted as 'F') connected by rapid saccades (coloured yellow). It is seen that gaze movement is neither regular nor uniform: at times the gaze location moves right across the picture while, at other times, it moves to a nearby feature. Also, certain features of the picture—human faces, for example—receive more of the user's attention than the background architectural structures and the table.

Saccades are very rapid movements of the eyeball, attaining an effective angular rotation of up to 600°/second (Becker 1991). As pointed out in the discussion of saccadic blindness in Chap. 2, the viewer is momentarily effectively "blind" during these movements, although, as with blinking, there is no awareness of this in normal vision. Saccades are "pre-determined": gaze jumps automatically to a place determined by the brain's visual system during the preceding fixation(s).

---

[1] Tatler and Wade (2003) give a description and history of various types of eye movements; Findlay and Gilchrist (2003) discuss their broader context in human vision.

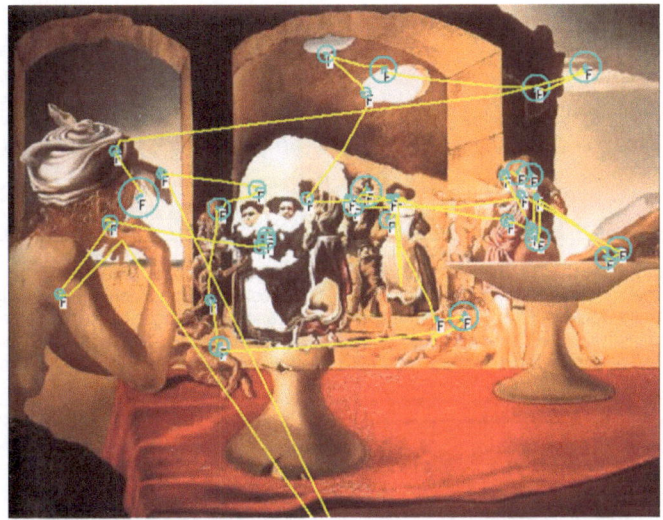

Fig. 4.1 Typical pattern of fixations and saccades when viewing a picture

There is a limit to how far gaze can be shifted during a saccade and if a movement of more than about 15° is required, the head will also rotate automatically.

The software used to represent the fixation and saccadic information in Fig. 4.1 has identified each fixation with an 'F' and a cyan point and circle, the diameter of the latter being proportional to the duration of a fixation. Fixation durations vary considerably. In the 12.5 s recording there are 38 fixations of average length 274 ms, accounting for about 83 % of the total time. For comparison with the other three gaze strategies to be introduced, this visual search strategy is summarised diagrammatically in Fig. 4.2a.

Note that certain commonplace activities, such as reading, also give rise to distinctive gaze patterns. Reading, for instance, is characterised by an ordered sequence of fixations along each line of text, though generally not equating to recognised word boundaries (e.g. McConkie 1983; Rayner 1998). A variety of different gaze strategies relevant to RSVP are shown in Fig. 4.2 and are described in the sections that follow.

4.2.2 Steady Gaze

If images are presented at a typical RSVP rate of 10 per second there is no time for gaze to move around an image in a characteristic fixation + saccade pattern; rather, there is a tendency for gaze movements to cease, often becoming centralised in a relevant location for extended periods of time. We call this behaviour steady gaze, summarised diagrammatically in Fig. 4.2b.

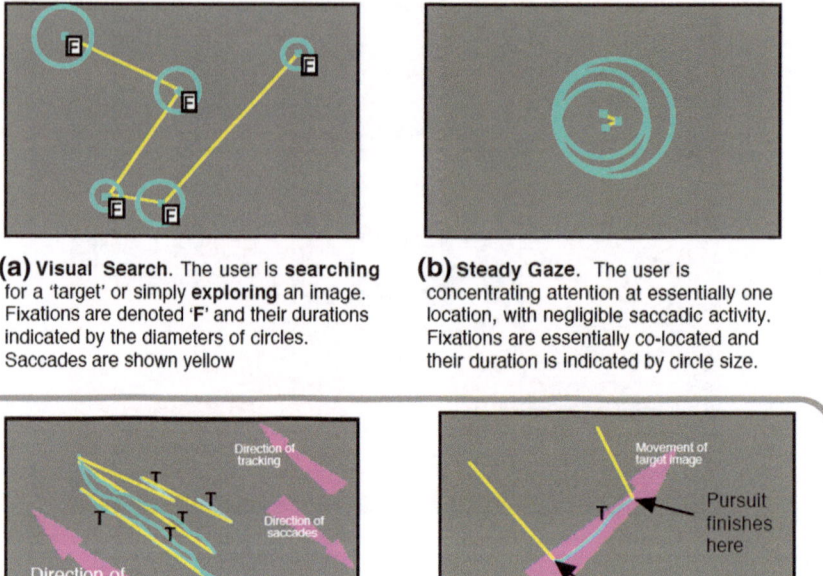

(a) Visual Search. The user is **searching** for a 'target' or simply **exploring** an image. Fixations are denoted 'F' and their durations indicated by the diameters of circles. Saccades are shown yellow

(b) Steady Gaze. The user is concentrating attention at essentially one location, with negligible saccadic activity. Fixations are essentially co-located and their duration is indicated by circle size.

(c) Nystagmus. Gaze tries to 'keep up' with the continuous stream of images. T indicates tracking.

(d) Visual pursuit: Following a saccade, gaze tracks a (possible) target image

Visual tracking

Fig. 4.2 Diagrammatic representations of the four gaze strategies. **a** Visual search. **b** Steady gaze. **c** Nystagmus. **d** Visual pursuit

Steady gaze can be distinguished from the intentional act of "staring" at something, in that the user adopts the steady gaze pattern without any voluntary thought or intention. It has been suggested that the observed improvements in text RSVP are largely due to the absence of saccades due to the single viewing location (Rubin and Turano 1992), although this is unlikely to account for all the improvements reported.

4.2.3 Nystagmus Gaze Tracking

Not all image presentations are static: in many of the RSVP modes we have encountered images follow defined paths at a range of speeds. Accordingly, human vision adapts to accommodate to these circumstances. Of particular relevance to gaze behaviour are adaptations to the movement of either images on a display or physical movements by a person—the phenomena called *nystagmus* and *visual pursuit* are two variants of what is generally termed 'visual tracking'.

Nystagmus can perhaps best be explained by reference to a common experience, that of looking out of the window of a fast moving train and attempting to 'keep up' with the view rather than examine some feature within that view. The user is trying to stabilise the larger part of the visual field. Similarly, when observing rapidly moving images in an RSVP mode the eye will also adopt these characteristic nystagmus[2] movements. When viewed as angular movement with time, these movements have a distinctive repetitive "sawtooth" motion, a rapid ramp of movement followed by an even more rapid (saccadic) return. An illustrative diagrammatic representation of nystagmus is shown in Fig. 4.2c, and will be discussed in detail later in the context of moving RSVP mode designs (Sect. 5.2).

4.2.4 Visual Pursuit

While nystagmus is a response to generalised rapid movement in the field of vision, gaze response to specific moving objects is different. In these circumstances gaze may be seen to follow an object, much as a photographer might swing his camera to follow a moving object, keeping it sharp at the expense of a blurred background. We refer to this gaze behaviour as *visual pursuit*.

A diagrammatic representation of a visual pursuit event in response to a moving image can be seen in Fig. 4.2d. Visual pursuit will be discussed in detail in the context of multiple entry/exit RSVP modes in Sect. 5.3. There are limits to the speed of movement that can be accommodated by these forms of active gaze tracking and we consider this later. The gaze analysis software used can isolate either type of tracking event and shows them as an elongated cyan trace, marked with a 'T' to distinguish them from fixations ('F').

4.3 Structure of the Eye

Two questions immediately arise from the evidence presented in Fig. 4.1 above. First, why does gaze have to be directed between a number of locations in a scene? Why can't the user simply take one 'look' at a scene and immediately know it in all its detail?

The answer lies in the make-up of the retina (Fig. 4.3, right) at the back of a user's eye (Fig. 4.3, left). A relatively small and approximately circular area called the *fovea* is made up of a high density of cones (which provide colour sensitivity)

---

[2] These entirely normal gaze movements should not be confused with the medical condition also called *nystagmus*, where the eyes make continuous uncontrolled flicking movements and which has highly detrimental effects on vision: (http://en.wikipedia.org/wiki/Nystagmus).

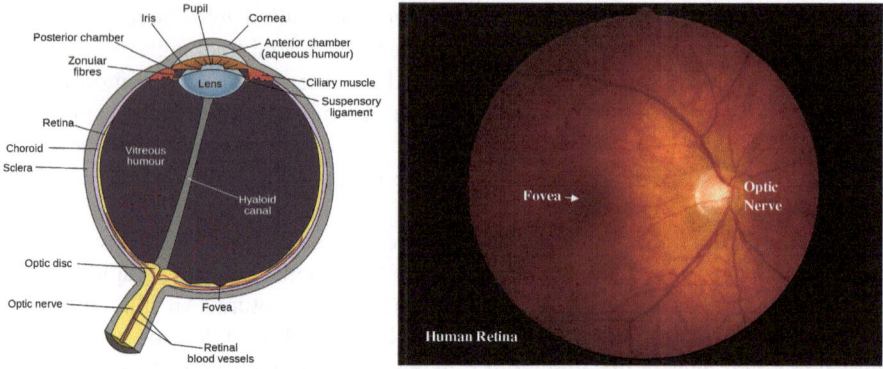

Fig. 4.3 (*left*) Anatomy of the human eye, (*right*) human retina

but a low density of rods, which do not respond to colours. The high density of cones provides extremely high acuity and sensitivity to colour within the visual angle subtended by the fovea (about one degree). The effect (and purpose) of fixations is to move specific parts of the visual scene to be within this high acuity foveal area, where detailed visual processing is possible.

Movement of the eye within its orbit is controlled by six muscles. These muscles are capable of the rapid and controlled movements of the eye, up to 600°/sec for saccades and around 100°/sec for tracking events. Equally they hold the eye stationary during fixations.

Away from the central fovea rod cells predominate. They are far more sensitive to light than cones and they are more numerous in the retina away from the fovea, but are not sensitive to colour (Fig. 4.4). A sparser covering of cones in this peripheral region maintains some colour perception. The ability to recognise detail drops off rapidly away from the central fovea.[3] As a general principle, objects must appear larger or be more distinctive to be accurately identified towards the edge of the visual field.

The effective size of the fovea is illustrated in Fig. 4.5. The angle subtended by the foveal region of the retina is around one degree, so at a distance of one metre the area of a distant display seen in great detail is about 1–1.5 cms in diameter: if you extend your arm, the width of the thumb nail is a reasonable indicator of the foveal angle. The area immediately around the fovea and subtending around 5° (roughly the width of a hand at arm's length) is called the *extrafovea* or *parafoveal region*: outside that we talk about *peripheral vision*. The display of

---

[3] It's true—try this perceptual party trick. Seat yourself in a normally lit room and stare continuously at a fixed point on the opposite wall. Ask a friend to move a suitable object, say a playing card, gradually towards your fixed line of sight from a few metres away. You will be surprised how close the card has to be to your sightline before you can reliably call colour, suit or rank. Of course, if you cheat and glace at the card—even for an instant—recognition is easy. See also Anstis (1974).

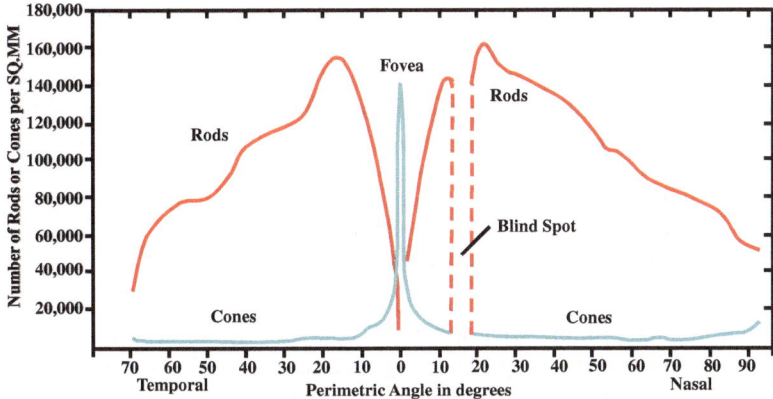

Fig. 4.4 Distribution of rods and cone cells across the fovea (adapted from Osterberg 1935)

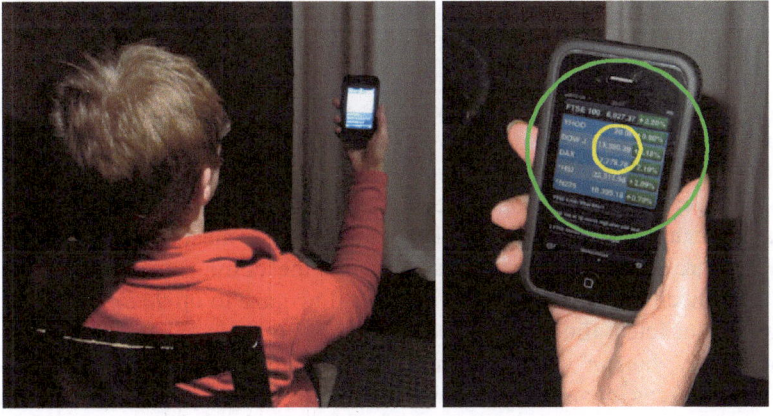

Fig. 4.5 Viewing a mobile device—approximate foveal (*yellow*) and parafovea (*green*) areas

a mobile or PDA held at arm's length falls mostly within the foveal region and completely within the parafovea, as illustrated in Fig. 4.5 respectively as yellow and green circles.

If we could only 'notice' items in the foveal region the number of car accidents would be much higher than it is. While the detail of a static image in the peripheral region cannot be deciphered in anything like the detail associated with foveal vision, *movement* in the peripheral region can, for example, easily be noticed. So can the change in colour of a traffic light. Nevertheless, the foveal angle is sufficiently small as to require the gaze behaviour illustrated in Fig. 4.1. Indeed, it may be useful to point out that, as far as the eye is concerned, the sequence of images projected onto the retina in the example of Fig. 4.1 could be considered to be a Rapid Serial Visual Presentation, especially since no rapid image movement is detected during a saccade. The question we now need to ask is what the user is actually *doing*, perceptually and cognitively, during fixations and saccades, and to explore the notions of salience and attention that would seem to lie at the heart of the task that a user is undertaking.

4.4 Where does the Eye Look Next?

Gaze control—determining what a person looks at—varies according to the context in which looking is taking place (e.g. Yarbus 1967). What the person attends to and their eyes fixate on is completely different when driving (road conditions, hazards, signposts and so on), as to when one is a passenger. If this were not so then car travel, and life in general, would be substantially more hazardous than it is. The immediate target of gaze control is subject to instant and conscious modification, over-riding the brain's autonomous choices. Nevertheless, there will be topics of continuing interest to individuals, which persist and may become evident under otherwise controlled test conditions.

There is a strong trade-off between angular distance from the fovea and image size when identifying images. As we have seen, detail is lost in the periphery of vision: an image easily identified when near the fovea may be completely unrecognisable in peripheral vision. The relationship between visual acuity and distance from the fovea is neatly encapsulated by Anstis (1974).[4] This trade-off is particularly significant for RSVP modes where small images must be accurately reported but which are located, or appear, away from the current point of gaze.[5] To be effective a mechanism must be provided to allow gaze to move to and fixate on an image, and sufficient time must be permitted to allow this to happen.

Fortunately, human vision is remarkably sensitive to general shapes and outlines, colourations, changes and movements in the visual periphery. Generally the appearance of a significant image will attract visual attention and the gaze point will move to it at the next or subsequent saccade. However, recall the problems of "cognitive blindness" (introduced in Sect. 2.5) caused by visual distraction or masking. These are of real concern in RSVP application design. If several aspects in the presentation change simultaneously, or even in quick succession, one of these changes may suppress recognition of the desired image—a manifestation of the "mud splash" effect described in Sect. 2.5.2.

There is an ordering or salience to items and images in the visual field, and this appears to determine what is attended to and which attracts gaze, and what is passed over. As the designer can rarely control the effective salience of images, care must

---

[4] Recall the game described in footnote 3.

[5] This effect, as we shall see, is particularly apparent in the Collage mode design investigated in Sect. 5.3.6.

be taken in an application design to ensure that items, which may not naturally be conspicuous, are not masked by potentially more noticeable and distracting images.

There is no one clear model that represents visual salience across all its properties, but several salience and eye-gaze models have been proposed. Itti et al. (1998) present a computer model of "bottom-up" (image data) driven salience, in which features (based on properties such as intensity, edge orientation, colour, etc.) are extracted from the image at multiple scales and salience assigned according to the application under consideration. Stark and Choi (1996) present a model that emulates the eye-gaze path of a simulated human observer using a Markov state model based approach. Witkowski and Randell (2007) present a model based on object and task precedence to account for rapid shifts in gaze strategy.

4.5 Detecting and Recording Gaze

A typical set-up[6] that detects eye-gaze is shown in Fig. 4.6 (left). An infra-red LED mounted in a camera lens directs its beam safely into the human eye. The camera, mounted directly below the display screen in Fig. 4.6 (centre picture), records the infra-red light reflected from both the retina and the surface of the cornea (shown in Fig. 4.6, right). A calculation based on the relative positions of the larger retinal reflection (through the pupil)[7] and the smaller corneal reflection from the surface of the eye can then determine to within an accuracy of about 2 mm the place on the display at which the user is looking.

Typically, the location of eye-gaze on the display is recorded at a rate of 50 or 60 per second (depending on the make of equipment used) and the system must be calibrated for each user before recordings can be made. Blinks—which typically

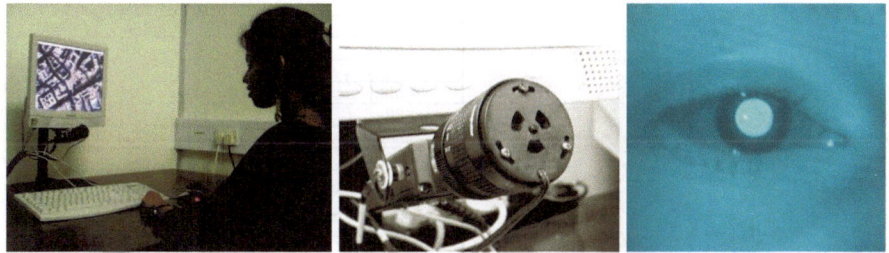

Fig. 4.6 (*left*) Eye-gaze recording set-up, (*centre*) gaze camera, (*right*) retinal and corneal reflections

---

[6] An early LC Technologies, Inc. system, see http://www.eyegaze.com/. Some of the data we present was captured on a Tobii system (http://www.tobii.com/).

[7] Similar to the "red-eye" effect often seen in flash photographs of faces.

last around 100 ms (equivalent to five or six gaze readings)—are also recorded. A full description of gaze recording technologies and techniques can be found in Duchowski (2003).

4.6 Representing and Analysing Gaze Recordings

The question arises as to how best to represent highly dynamic gaze data. The gaze recording equipment simply generates a long sequence of X-Y coordinates of gaze position on the display, once every 16.7 or 20 ms (60 or 50 times a second). The first task is to extract fixations. This is done automatically in software. By convention, a fixation is usually defined as a run of six or more gaze coordinate readings where the gaze location on the screen has not varied by more than a small distance between any of the readings.[8] We generally use a distance equal to a capture radius of 12 pixels, though individuals differ in the natural instability of their gaze (sometimes called eye "jitter"): increasing the capture radius can compensate for this. Once a single gaze reading falls outside the radius criterion the fixation is considered finished. This is usually due to the start of a saccade.

By changing the fixation detection algorithm to allow the fixation to continue while each successive reading (as opposed to *all* readings) remains within the current capture radius criterion, tracking events are effectively captured, as observed in both nystagmic behaviour and visual pursuit (Sect. 4.2.3). Again a single reading outside the capture radius signals the end of the tracking event. Tracking detection is an option in the analysis software, and is used extensively in analysing the moving and multiple entry/exit modes later.

Figure 4.7 (left) shows a short (3.25 s) extract from the gaze trace shown in Fig. 4.1, and a tabular form of the fixations extracted from the sequence is shown to the right. In this case the analysis software has been used to

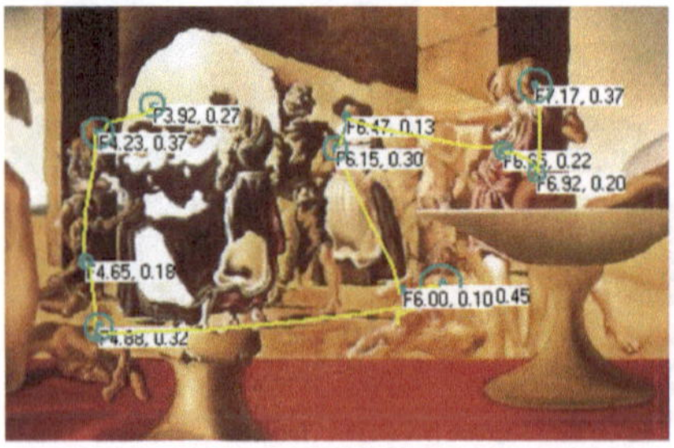

Start	Duration	Samples	X mean	Y mean
3.917	0.267	16	400.4	277.9
4.234	0.366	19	344.3	306.2
4.650	0.184	11	329.6	443.5
4.884	0.316	19	343.2	513.1
5.517	0.450	27	704.5	468.3
6.000	0.100	6	664.3	472.7
6.150	0.300	18	593.6	322.7
6.467	0.133	8	605.8	288.6
6.650	0.217	13	768.2	324.6
6.917	0.200	12	804.8	349.1
7.167	0.367	22	802.5	256

Fig. 4.7 (*left*) Gaze track of saccades and fixations, detail, (*right*) tabulated fixation data

---

[8] Actually there are a number of different measures to define a fixation (e.g. Salvucci and Goldberg 2000), but in practice the differences between them are not significant for the type of analysis we are conducting here.

additionally show the fixation start times (in seconds from the beginning of the recording) and durations (seconds) on the trace. This form of representation is very useful for short sequences and details of particular segments of gaze behaviour, but quickly becomes cluttered as more of the gaze trace is included.

4.6.1 Temporal Representation

We use an XY-T plot to graph the X and Y position of the gaze point on the display as a function of time. Figure 4.8 shows this gaze representation method for the case of a visual search. The X (horizontal) coordinates on the display are shown in red and the Y (vertical) coordinates in blue. In these XY-T plots the top-left coordinates of the display (Y = 1, X = 1) are both shown at the top of the graph. Fixations appear as plateaux in both X and Y, and saccades as abrupt vertical changes. Any blinks would appear as short interruptions to the plot. Temporal representation of RSVP was pioneered by de Bruijn and Spence (2002).

Figure 4.9 shows an example of an XY-T plot for a steady gaze example (refer back to Fig. 4.2b and Sect. 4.2.2). It can be seen that the X and Y coordinates

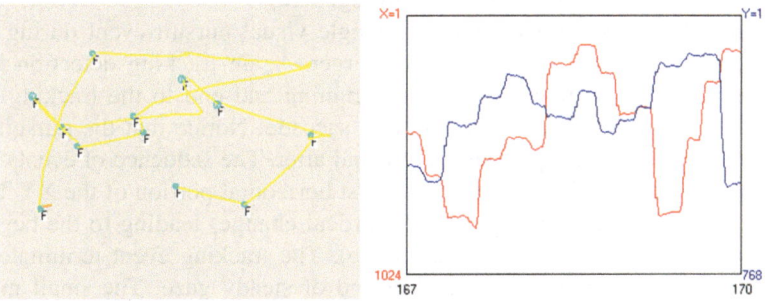

Fig. 4.8 XY-T plot representation showing a pattern of fixations and saccades

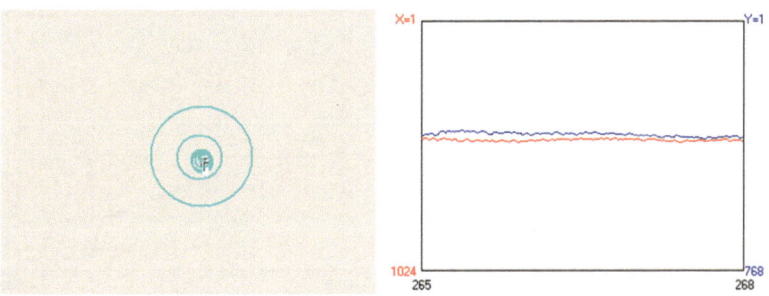

Fig. 4.9 Gaze track and XY-T plot for steady gaze

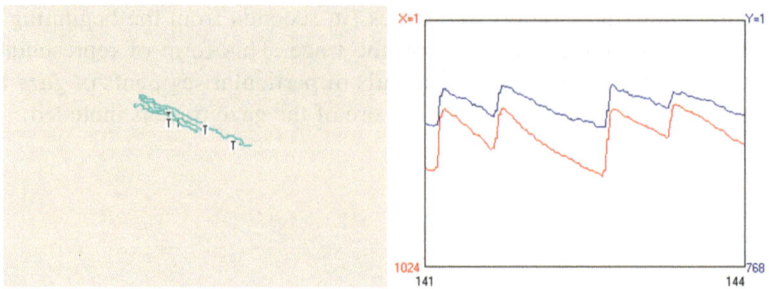

Fig. 4.10 Gaze track and corresponding XY-T plot for nystagmus

of the gaze location on the screen are stable. Slight movements in the gaze point cause the fixation algorithm to register several distinct fixations.

Temporal representation can also be valuable in clarifying the motion of the gaze point on the display in highly dynamic situations. Figure 4.10 shows an XY-T plot for a nystagmus mode sequence. Note the distinct "sawtooth" appearance of the plot. Tracking detection has been used in this example to highlight the smooth movement of the gaze point (marked with a 'T') as it follows the general direction of motion on the display. For clarity, the yellow saccadic movements are omitted from the gaze track image (compare with Fig. 4.2c).

Figure 4.11 shows an example of a single visual pursuit event during which gaze closely follows a specific moving target. Again tracking detection is enabled. The XY-T plot shows a distinct beginning and end to the tracking movement, starting and ending with a (yellow) saccade. Notice that the pursuit event is flanked by steady gaze periods, before and after. The sequence of events shown begins with a period of steady gaze (the first horizontal portion of the XY-T plot), which ends with a saccade (the abrupt vertical change) leading to the beginning of the tracking event (the downward slope). The tracking event terminates with another saccade leading to a second period of steady gaze. The small magenta segment embedded within the cyan tracking event is used to indicate a user response.

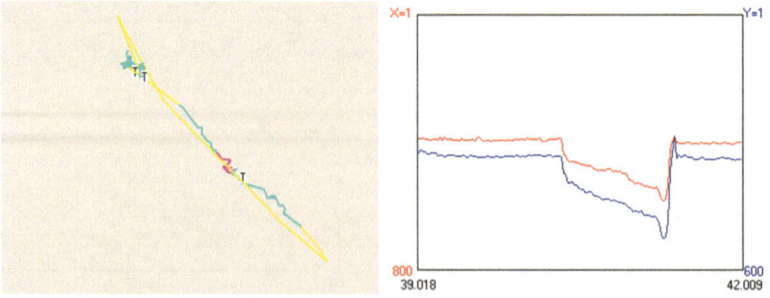

Fig. 4.11 Gaze track and corresponding XY-T plot for visual pursuit

4.6.2 Heatmap Representation

Another useful representation of gaze behaviour—the "cumulative heatmap"—is shown in Fig. 4.12. Heatmap representations are particularly useful for getting an overall idea of gaze activity across many trials or samples. Colours represent the density of gaze activity at locations on the display over an extended period of viewing or accumulated across several users.

Compare, for instance, the "raw" gaze track, mostly saccades shown in yellow, for a user viewing Volcano mode for 223 s (Fig. 4.12, left), with the equivalent cumulative heatmap shown in Fig. 4.12, right. Both show the same trial data of over 11,000 individual gaze point readings. In the heatmap the highest levels of gaze activity are shown in red, while blue represents an absence of activity, with intermediate levels of activity coloured according to the scale shown to the right. All the heatmaps shown are scaled and coloured in this way. When comparing like recordings, a heatmap with large red areas does not imply more activity overall, only that the gaze activity is more evenly distributed over the display.

4.6.3 Gaze Travel

One further useful measure of gaze activity is "gaze travel". In this measure the total accumulated distance travelled by the gaze point across the display is calculated as the sum of the Euclidian distances between successive gaze point recordings. As the distance travelled during saccades is much higher than during fixations, gaze travel represents a useful single measure of how extensive gaze movement was during a period of recording. We will usually present this distance in "screen pixels". A related measure to indicate gaze activity is that of gaze speed, expressed in screen pixels per second (pix/s). Where the geometry of the display and viewing distance are known, gaze travel can be expressed as effective angular travel or angular gaze speed of the eyeball (degrees or °/s).

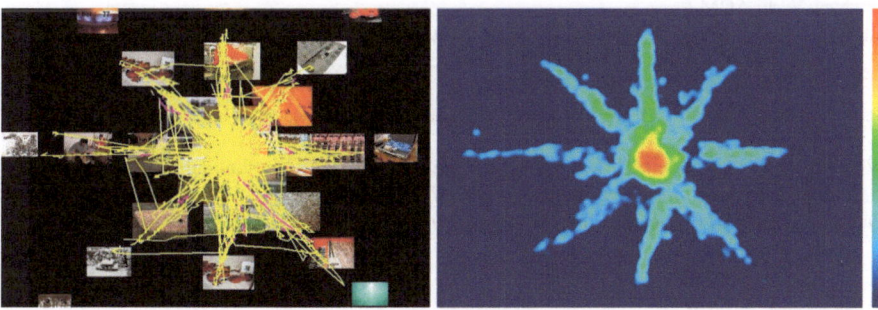

Fig. 4.12   Volcano mode: (*left*) gaze track and (*right*) equivalent heat map representation

Gaze travel measures are useful for comparing like recordings. For instance a presentation where static gaze predominates will exhibit a low gaze travel distance, while, for example, a nystagmic one will be characterised by very high travel distance.

In the next chapter we present an analysis of user gaze behaviour for ten of the RSVP modes described in earlier chapters: three "static" modes, three "moving" modes and four "multiple entry/exit" modes. We will see that gaze behaviours vary widely according to the RSVP mode and that the gaze strategy will adapt naturally according to the task the user has been set.

References

Anstis, S. M. (1974). A chart demonstrating variations in acuity with retinal position. *Vision Research, 14*, 589–592.

Becker, W. (1991). Saccades. In R. H. S. Carpenter (Ed.), *Eye movements* (Vol. 8, pp. 95–137), Vision and visual dysfunction Boca Raton: CRC Press.

Cooper, K., de Bruijn, O., Spence, R., & Witkowski, M. (2006). A comparison of static and moving presentation modes for image collections (pp. 381–388). *Proceedings of Advanced Visual Interfaces (AVI-2006)*.

de Bruijn, O., & Spence, R. (2002). Patterns of eye gaze during rapid serial visual presentation (p. 11). *Proceedings of AVI-02*.

Duchowski, A. T. (2003). *Eye tracking methodology: Theory and practice*. New York: Springer.

Findlay, J.M., & Gilchrist, I.D. (2003). Active vision: The psychology of looking and seeing. Oxford: Oxford University Press.

Itti, L., Koch, C., & Niebur, E. (1998). A model of saliency-based visual attention for rapid scene analysis. *IEEE Trans Pattern Analysis and Machine Intelligence, 20*, 1273–1276.

McConkie, G. W. (1983). Eye movements and perception during reading. In K. Rayner (Ed.), *Eye movements in reading* (pp. 65–96). New York: Academic Press.

Osterberg, G. (1935). Topography of the layer of rods and cones in the human retina. *Acta Ophthalmologica, 13*(6), 11–103.

Rayner, K. (1998). Eye movements in reading and information processing: 20 years of research. *Psychological Bulletin, 124*(3), 372–422.

Rubin, G. S., & Turano, K. (1992). Reading without saccadic eye movements. *Vision Research, 32*(5), 895–902.

Salvucci, D.D., & Goldberg, J.H. (2000). Identifying fixations and saccades in eye-tracking protocols (pp. 71–78). *Proceedings of the Eye Tracking Research and Applications Symposium*. New York: ACM Press.

Stark, L. W., & Choi, Y. S. (1996). Experimental metaphysics: The scanpath as an epistemological mechanism. In W. H. Zangemeister, et al. (Eds.), *Visual attention and cognition* (pp. 3–69). Amsterdam: Elsevier.

Tatler, B. W., & Wade, N. J. (2003). On nystagmus, saccades and fixations. *Perception, 32*, 167–184.

Witkowski, M., & Randell, D. A. (2007). A model of modes of attention and inattention for artificial perception. *Bioinspiration and Biomimetics, 2*, S94–S115.

Yarbus, A.L. (1967). Eye movements and vision. New York: Plenum (Originally published in Russian 1962).

Chapter 5
Analysing Gaze for RSVP

Abstract In this chapter we analyse 10 distinct RVSP modes drawn from across the whole range of static, moving and multiple entry/exit modes in terms of the visual search, steady gaze, nystagmus and visual pursuit characterisations introduced in the previous chapter. Eye-gaze response behaviour is analysed with quantitative data and fully illustrated with gaze plots of fixations and saccades, temporal plots of the x and y coordinates of gaze position (XY-T), and cumulative gaze activity heatmaps. Consideration is given to the reason why, at any instant, gaze might move to a new position and the influence this might have on RSVP design.

Keywords Eye-gaze behaviour • Variations in gaze behaviour • Static RSVP modes • Moving RSVP modes • Multiple entry/exit modes • Fixation plots • XY-T plots • Cumulative gaze heatmap plots

By means of the analytical tools and representations developed in the last chapter we now examine ten common RSVP modes selected from the three classes of static, moving and multiple entry/exit. We do so in order both to understand how gaze behaviour is influenced by the nature of each mode as well as to provide guidelines to an interaction designer who wishes to exploit these and other modes in an application.

5.1 Static Presentation Modes

First, we consider the difference in gaze behaviour between Slide-show mode RSVP, in which the user gets a single brief view of each image in a set and must recognise a specific image, and Tile mode, where all the images in the set are visible simultaneously. An interesting third example, Mixed (2 × 2) mode, combines aspects of both Slide-show and Tile mode. The gaze data used is that from the study by Cooper et al. (2006).

5.1.1 Slide-Show

It is immediately clear when analysing gaze behaviour patterns associated with Slide-show mode that the gaze point remains essentially fixed at the centre of the display area for the entire duration of the presentation. This is in direct contrast to the gaze behaviour that would be expected if the user were presented with each image individually for an extended period, as in Fig. 4.1, when a series of saccades and fixations around the image area would be expected. With only a short image exposure, we may presume that the user is reliant on rapid image "gist" recognition. Additionally, one might reasonably conjecture that centring on the image maximises the coverage of the higher resolution areas of the retina.

Based on gaze data from 10 users, Fig. 5.1 shows the cumulative heatmap gaze pattern for Slide-show mode, confirming the pattern of centralised steady gaze behaviour in this presentation mode. Not all individuals select the exact centre and Fig. 5.2 illustrates some of that variation.

Figure 5.2 (left) from participant #5 shows 5 fixations, all nearly closely co-located (average length 1,097 ms, longest 1,670 ms), whereas Fig. 5.2 (right) from participant #7 shows 12 rather more widely spread fixations in the same exposure time (average fixation length 418 ms, longest 850 ms). In the latter example the fixations are both

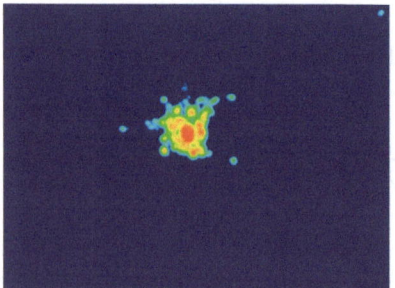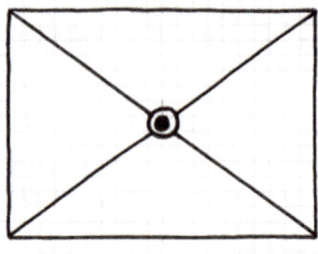

Fig. 5.1 Slide-show mode: cumulative heatmap (*left*), notation (*right*)

Fig. 5.2 Slide-show mode fixation pattern examples showing the natural variability between individuals

shorter and more scattered about the central point. This illustrates the difference in gaze stability between individuals—although there is no change in overall gaze strategy for this presentation mode from any tested individual.

5.1.2 Tile Mode

Tile mode (Fig. 5.3) presents all the images in the set, including any target images, simultaneously. In complete contrast to Slide-show mode, gaze behaviour adopts the visual search pattern characterised by short fixations interspersed by rapid saccades. Even though an explicit search task is being undertaken, this pattern of gaze activity is very similar to that observed when a user is simply viewing a single image of equivalent size.

The cumulative heatmap (Fig. 5.3, left) shows a generally even distribution of gaze over the image area. The few "hotspots" relate to apparently distinctive images (and target areas) that are visited more than others. Little is known about the mechanism by which the pattern of search is controlled, but it is almost never systematic or evenly distributed over the available search area.

Figure 5.4 (left) shows an illustrative example of gaze behaviour for Tile mode. During a presentation lasting 2.9 s, 11 fixations of average length 235 ms (longest 570 ms) were recorded. Figure 5.4 (right) shows the corresponding XY-T plot

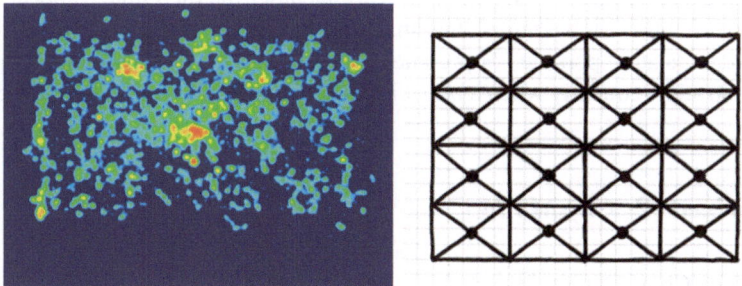

Fig. 5.3   Tile mode cumulative heatmap (*left*), notation (*right*), not to scale

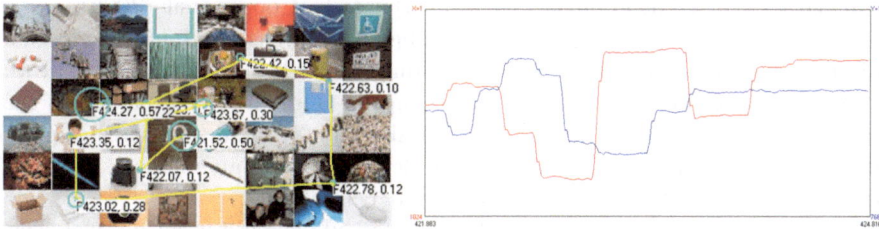

Fig. 5.4   Tile mode single exposure (2.9 s) (*left*), XY-T plot (*right*)

Fig. 5.5 Search to target (*left*), pre-attentive discovery (*right*)

representation. The pattern of saccades (steps) and fixations (plateaux) can clearly be seen. The target was not present in this example.

When a target image is present in Tile mode, it is sometimes detected after a period of visual search and sometimes immediately ("pre-attentively"). Figure 5.5 (left) shows a period of visual search prior to locating the target—the picture of a human eye in the rightmost column—with four distinct saccades and fixations before the target is acquired. The brief fixation just prior to a small "correction" onto the main target is relatively common after a long saccade and probably indicates a "falling short" by the ballistic saccadic estimation process (see Becker 1991, for a detailed description of the mechanism of saccades).

In contrast, the Tile mode trace shown in Fig. 5.5 (right) illustrates immediate acquisition of the target: the first complete saccade after the tiled image appears leads directly to a fixation on the target (the image of orange circle shapes). This example, we believe, indicates "pop-out" or pre-attentive recognition as discussed in Chap. 2. It occurs relatively frequently when a target is visually distinctive within the image set.

5.1.3 Mixed (2 × 2) Mode

Recall from Chap. 2 that Mixed mode was intended to provide longer viewing times for images, with a relatively minor, but compensating, reduction in size. Figure 5.6 (left) shows the cumulative heatmap for a range of examples of Mixed 2 × 2 mode, combined over the three different overall presentation rates (3.57, 2.50 and 1.92 mixed quad images per second) used by Cooper et al. (2006).

Figure 5.7 shows two illustrative examples of higher and lower presentation rates in 2 × 2 Mixed mode. With a longer overall presentation time (lower presentation rate) the user's gaze typically moves in a non-regular pattern of saccades and fixations spread around the central region of the display area, Fig. 5.7 (left). As the presentation time decreases (rate correspondingly increases), Fig. 5.7 (right), the fixations become increasingly more centralised.

Overall this trend is consistent with an observed decrease in gaze travel speed with increase in presentation rate: 739 pix/s for a rate of 1.92 quad presentations/

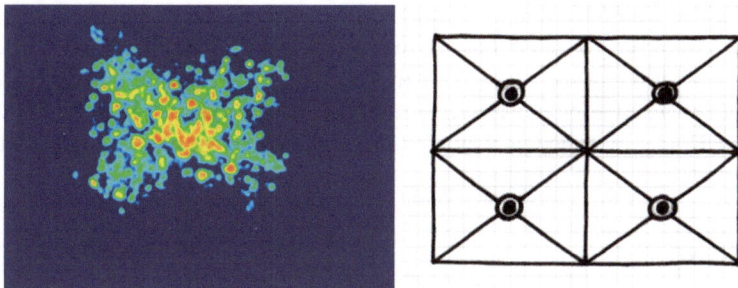

Fig. 5.6   Mixed 2 × 2 mode cumulative heatmap (*left*), notation (*right*)

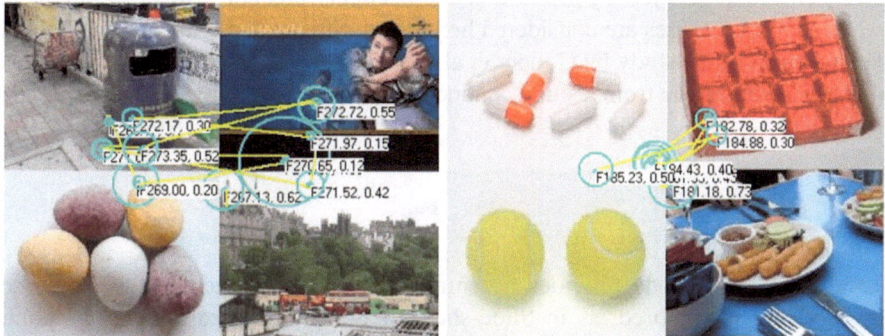

Fig. 5.7   Slower presentation rate (*left*), higher presentation rate (*right*)

second (i.e. the 12 quad images in 6.24 s), as opposed to a gaze travel of 430 pix/s for a rate of 3.57 quad presentations/second (i.e. the 12 quad images in 3.36 s). It is not established whether very high presentation rates, comparable to that used in Slide-show mode, would give rise to fully centralised gaze.

5.1.4 Summary

Slide-show mode RSVP invariably evoked a steady gaze response. This response is similar to that observed for text RSVP and gives rise to low values of gaze travel; it is generally well liked by users and effective as a method of presentation. At high presentation rates blinking becomes a potential problem, as the perceptual blindness caused can completely mask a target image appearance.

Tile mode gave rise to a distinctive visual search strategy of saccades and fixations, as observed when viewing a normal picture on a display. We noted that in some instances the target image was noticed immediately (pre-attentively) and sometimes as a result of active search.

Mixed (2 × 2) mode, when presented comparatively slowly showed the characteristics of visual search mode, but becomes increasingly centralised and "steady" as image presentation rate increased. This mode was both successful, giving more time to view the images than Slide-show at a reasonable size, and well liked with low gaze travel rates. Due to extended presentation times, the effects of blinking are reduced.

5.2 Moving Modes

We now consider RSVP modes in which images have a substantive moving component. As with the static modes, changes in design have profound effects on the gaze behaviour adopted by users and these effects are not always as might be expected. Three modes are considered here: Ring, Stream and Diagonal mode, with data taken from the study by Cooper et al. (2006) and some additional observations about Diagonal mode from a study conducted with the University of Pavia in 2009.

5.2.1 Ring Mode

Recall that Ring mode provides a central area in which each image in the set appears, and is captured, as in Slide-show mode. Starting from the entry point, each image then moves in a spiral path leaving the display continuously at the top edge. The cumulative heatmap for this mode, Fig. 5.8 (left), indicates that the overwhelming majority of gaze activity is centralised and is of the steady gaze type. There are only very occasional excursions along the ring part, either following a very distinctive target or the last image as it leaves the screen.[1] Figure 5.9 shows an illustrative example; the average fixation length is 869 ms.

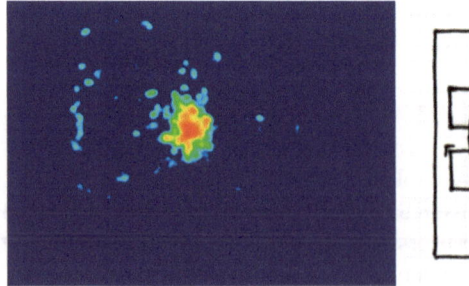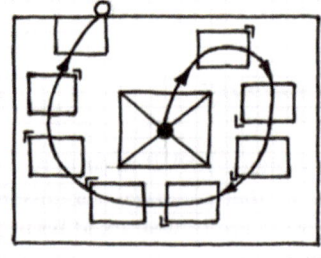

Fig. 5.8 Ring mode cumulative heatmap (*left*), notation (*right*)

---

[1] Cooper et al. (2006) report that at least one user adopted a nystagmatic gaze behaviour in this Ring mode—coupled to a dislike of the mode!

Fig. 5.9 Ring mode, illustrative example from one presentation

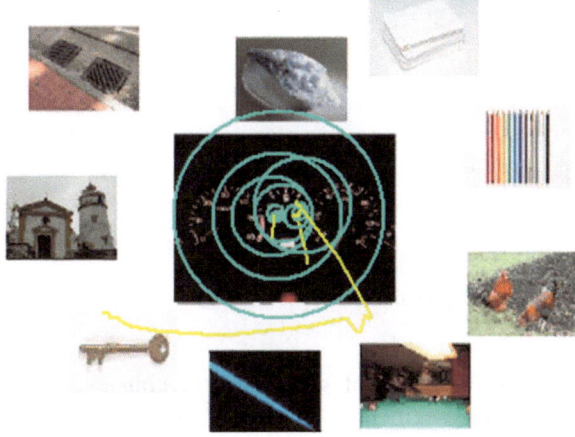

The clear implication of this analysis is that the moving component of this Ring mode design plays no effective part in the RSVP process and may be considered essentially decorative.

5.2.2 Stream Mode

Stream mode differs from Ring mode in that there is no capture point at any location. Instead the images are constantly moving, appearing continuously at the right and moving along a curved path to the top, where they disappear continuously. Figure 5.10 (left) shows a cumulative heatmap for this mode. There is a concentration of gaze activity along the portion of the image path where the images are at their largest. The central hotspot represents the gaze position at the start of each sequence and has no other significance.

More detailed inspection of the gaze path reveals that eye movement is nystagmatic. Figure 5.11 (left) shows an illustrative example. Here, instead of fixations,

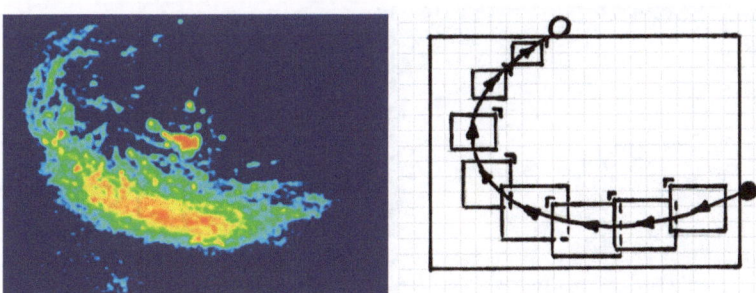

Fig. 5.10 Stream mode cumulative heatmap (*left*), notation (*right*)

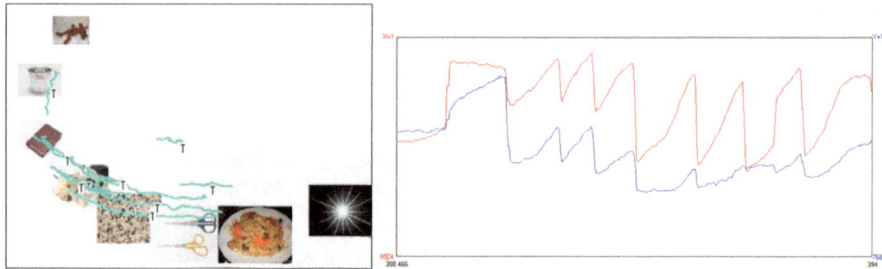

Fig. 5.11 Nystagmatic behaviour for one Stream mode sequence (*left*), XY-T plot (*right*)

a repetitive series of gaze path tracking events are observed (indicated with a "T"). Blue segments of the track indicate periods when successive gaze points are considered to comprise a single event. The usual yellow trace, representing the returning saccades, has been omitted for clarity. Figure 5.11 (right) shows the XY-T plot for the same sequence as Fig. 5.11 (left). The typical nystagmatic or sawtooth motion—ramp followed by rapid return—is characteristic of the image path tracking and saccadic returns just described. This gaze strategy—dominated by continuous gaze movements—is strongly linked to users' reported dislike of this Stream mode.

5.2.3 Diagonal Mode

In Cooper et al. (2006) Diagonal mode example images continuously appear at the top left corner and rapidly move diagonally across the display, leaving at the bottom right hand corner in a capture frame. This Diagonal mode design has many similarities to the Stream mode just considered; but is gaze behaviour equivalent?

The cumulative heatmap from 10 participants (Fig. 5.12, left) indicates two distinct types of response by individual users. Two separate areas of activity are

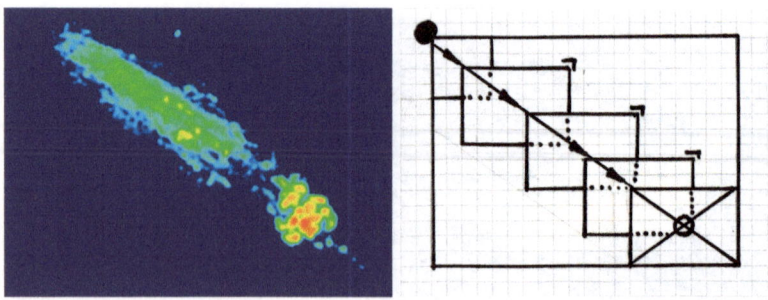

Fig. 5.12 Diagonal mode cumulative heatmap (10 participants) (*left*), notation (*right*)

apparent: first an elongated area diagonally across the screen, and second a group of locations near the exit point. In some cases, users will track the path of the moving image sequence nystagmatically in a manner very similar to the Stream mode response, but in about half of the cases gaze stabilises to a single steady gaze point near the bottom right capture frame.

Figure 5.13 (left) shows the response of one user: a series of closely spaced fixations near a single point. The associated XY-T plot confirms that this user moves from the (central) gaze point at the start of the sequence, follows the diagonal stream downwards and then remains steady near the bottom right corner (where the images are captured) until the end of the sequence. Figure 5.14 (left) shows the track for a single presentation sequence for a different user, showing the alternative nystagmatic strategy. This is confirmed by the XY-T plot (right).

Cooper et al. (2006) give no indication as to why users react differentially in this way in Diagonal mode, why it should be different from Stream mode, and whether there is any particular advantage gained by unconsciously adopting one or other of the gaze strategies. However, there are strong indications that users expressed greater preference for the Diagonal mode when they adopted the stable gaze strategy illustrated in Fig. 5.13.

In a separate study,[2] each user was presented with several instances of the Diagonal mode in succession (although with different paces and speeds) so that

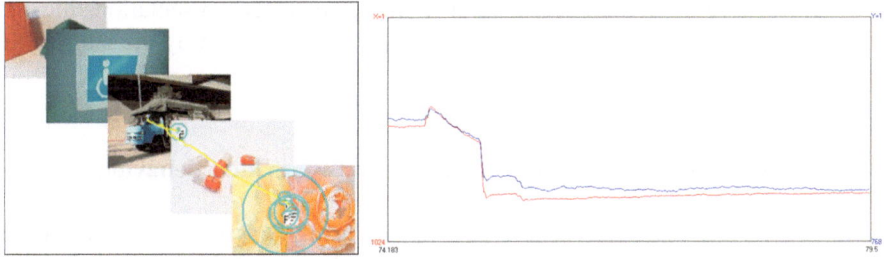

Fig. 5.13   Participant #10 with steady gaze (*left*), XY-T plot (*right*)

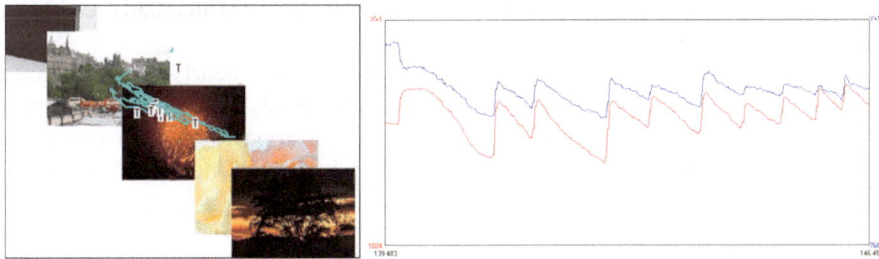

Fig. 5.14   Participant #11 showing nystagmus movement (*left*), XY-T plot (*right*)

---

[2] Conducted in collaboration with the University of Pavia.

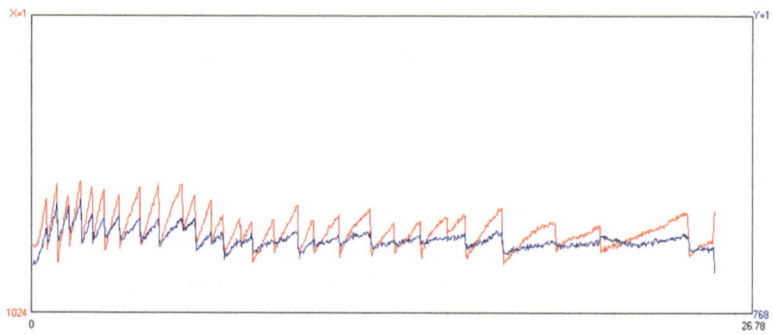

Fig. 5.15 Change from nystagmus towards stable gaze behaviour observed in diagonal mode

any change of strategy by an individual could easily be noted over the course of those presentations. Some users immediately demonstrated the steady gaze strategy (see Fig. 5.13) and some the nystagmus strategy (Fig. 5.14).

Much more clearly in the Pavia study than in Cooper et al's, there was a strong tendency for users to shift away from nystagmus tracking towards the steady gaze strategy over the course of the five individual presentations made to them. Once the steady gaze strategy had been adopted, there was no tendency to revert to the nystagmus strategy. Figure 5.15 shows a particular instance where the tracking was observed to slow markedly within a single sequence presentation, illustrative (but not typical) of the overall gaze strategy change shown by the majority of participants.

5.2.3.1 Effect of Image Speed and Limits to Gaze Tracking Behaviour

Figure 5.16 plots, for eight sample participants using Diagonal mode, the measured gaze speed across the screen against the actual speed of the image stream presented on the display. These are both expressed in pixels/second, and relate only to the X (horizontal) component of the gaze traces. The Y (vertical) component, however, demonstrates equivalent behaviour. Gaze speeds were manually estimated from the slope of the gaze traces.[3]

From this plot it generally appears that gaze speed nearly, but not quite, matches the image speed at lower presentation speeds, but does not exceed a gaze speed of approximately 1,000 pix/s. There appears to be a maximum angular speed for controlled gaze tracking, which will not be consistently

---

[3] We also performed a linear regression on several instances of gaze track data, but these computed values were not sufficiently different from those obtained by manual estimation to cause any concern.

Fig. 5.16 Gaze speed vs. image speed

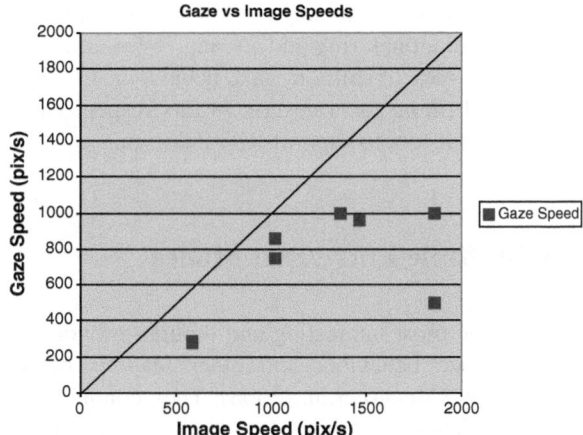

exceeded regardless of the speed of the image stream. While saccadic eye movements can demonstrate a much higher angular speed, they are ballistic and not controlled.

5.2.4 Comment

Ring mode, despite having a moving image stream, caused a steady gaze response similar to Slide-show mode. We conjectured, therefore, that the moving "ring" part was unlikely to play any significant role in the user's decision making process and should therefore be considered as an ornamental rather than functional aspect of this mode.

Stream mode, comprised only of moving images with no capture frame, evoked rapid and continuous eye movements (nystagmus), which were neither effective nor liked by the users.

Similarly, Diagonal mode had a high movement component, but did not always give rise to nystagmus—users naturally tending to either a steady gaze or a nystagmatic strategy. We also noted that users naturally tended to transition to a steady gaze strategy if they initially adopted the nystagmatic one. This tendency is apparently independent of the presence of a capture frame.

There will almost always be one particular concern in the mind of the interaction designer, and that is an application's visual appeal. While that cannot be quantified, it will nevertheless exert a strong influence on a design. Notwithstanding this concern with visual appeal, some clear advice can be drawn from the analyses just carried out. First, it appears that users favour a mode that allows them, for much of the time, to treat part of the presentation—that associated with a capture frame—as a Slide-show presentation. Useful gaze departures (i.e., other than nystagmatic) from such a location may be associated with context assessment or recognition confirmation, either or both of which may be beneficial to a specific application.

Similar conclusions may be drawn, and applied with caution, to moving modes other than diagonal, ring and stream, in which relatively small changes in a design can lead to radical shifts in gaze behaviour. In turn this will affect a user's performance and liking for a design. In this respect gaze recording and analysis represents a useful tool to support RSVP design.

5.3 Multiple Entry/Exit Modes

Perhaps the most interesting and thought provoking experimental results concerning the gaze behaviour and other features of multiple entry/exit modes was reported by Corsato et al. (2008). They conducted an experiment with five RSVP modes: Volcano, Floating, Shot, Collage and Grid.[4] The task given to subjects was a theme target recognition task: out of a total of 2000 images, as many of them relevant to a given theme (e.g., 'dog', 'cat', 'aeroplane') were to be identified out of a possible total of 40 within the image collection, though every user was unaware of the latter number. The recorded results included total gaze travel, the degree of fatigue reported by users and the identification score out of 40. These three measures are represented for the four modes also shown in Fig. 5.17. Subsequently, the gaze records were accessed to generate heat maps.

Some outcomes of the experiment immediately stand out. First, the target recognition score is noticeably higher for the Volcano and Floating modes than for the other modes. Second, these two modes are characterised by the lowest extent by far of gaze travel. Third, Volcano and Floating modes are associated with the least degree of fatigue reported by users. Fourth, heat maps show that, for Volcano and Floating modes, gaze is quite localised spatially.

At first sight Corsato et al's Volcano mode would appear to have features in common with Slide-show and Ring modes, as well as aspects of Diagonal mode. At the centre is a simple sequential presentation of the total image set, as in Ring mode, followed by the images leaving the screen along eight separate paths. The fact that there are eight paths from the centre, rather than a single path as in Diagonal mode, has the effect of considerably slowing the image movement and reducing the need for overlap. In fact, analysis of gaze behaviour for Volcano mode reveals that neither the Slide-show nor Diagonal modes on their own provide an adequate model for understanding gaze behaviour.

Floating mode resembles Volcano, except that the images are serially presented centrally at a much smaller size and grow in size as they move towards the edge of the display screen. Shot mode is distinct from Volcano and Floating modes. Images appear at a point and gradually increase in size along various paths down the display screen, leaving continuously at the bottom of the display area.

---

[4] Corsato et al's Grid mode is a close equivalent to Cooper et al's Tile mode and is omitted from consideration here.

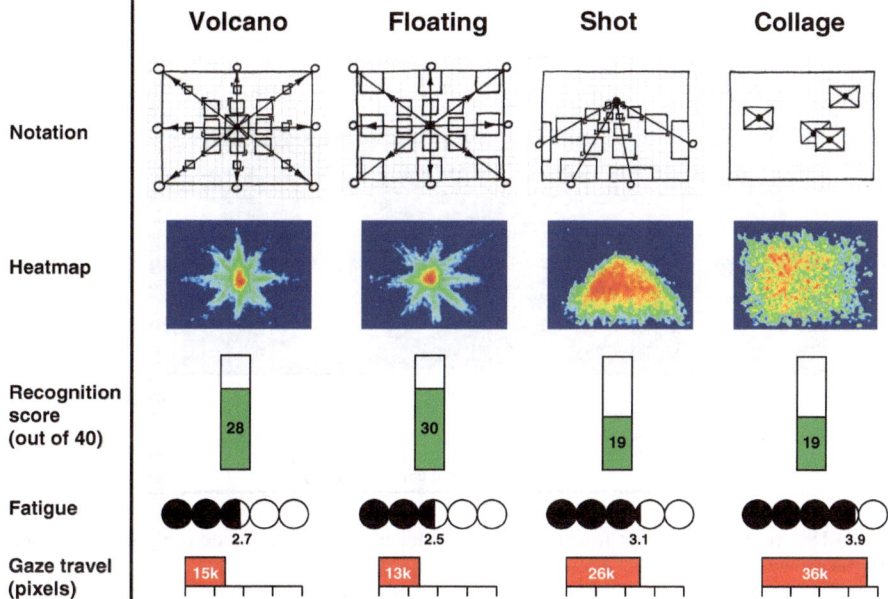

Fig. 5.17   Comparison of multiple entry/exit modes (data from Corsato et al. 2008)

Later we will investigate whether these characteristics of Shot mode gave rise to substantive differences in gaze behaviour.

5.3.1   Target Pursuit in Volcano, Floating and Shot Modes

Significantly, the gaze response of users to potential targets in these three multiple entry/exit modes is characterised by visual pursuit (refer back to Sect. 4.2.3) rather than the repetitive nystagmus pattern previously identified in Stream and Diagonal modes. In the three modes discussed in this section—Volcano, Floating and Shot—target appearance is strongly associated with individual visual pursuit events and these frequently result in target identification by user response. User responses are generally associated with correctly identified targets, but sometimes with a similar, but non-target image.

Figure 5.18 shows an example target pursuit event in Floating mode, in this case following the "dog" target image from the centre of the display to the right hand edge. The example is drawn from Floating mode, but gaze tracking follows a similar pattern in both Volcano and Shot modes. There is no immediate gaze response to the appearance of the image. Instead gaze can be seen to saccade (yellow) to the image as it moves along its designated path. Pursuit begins when visual tracking starts and ends with a distinct saccade. Where a target is identified

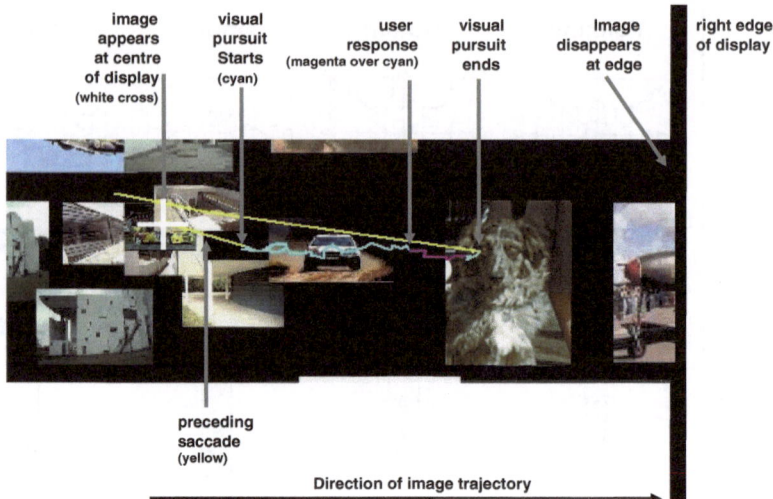

Fig. 5.18 An example target tracking visual pursuit in the floating mode (only part of the display is shown)

the user will indicate this by pressing a spacebar key (the user response). This is represented by a magenta coloured section of the overall cyan pursuit event.[5] The user response usually occurs before the end of gaze pursuit, but is sometimes delayed until after it is finished.

For the three RSVP modes Fig. 5.19 shows diagrammatically the average timings and relative changes in image size as the images move along their defined paths. The timings were extracted by individual inspection of a video recording of the presentation overlaid with the gaze data representation.

We next look at gaze behaviour for these three selected multiple entry/exit modes in individual detail.

5.3.2 Volcano Mode

A cumulative heatmap representing five complete user trials for Volcano mode is shown in Fig. 5.20. The preponderance of steady gaze activity at the central point of appearance of the image stream is clearly shown, as are the radial pursuit events along the eight directions used.

Some idea of relative times can be drawn from measurements of 54 instances of volcano use. On average (see Fig. 5.19) visual pursuit began 900 ms after the

---

[5] Only the start time of the user key press response is recorded in the experiment data, so the length of the magenta section is arbitrary and not significant here.

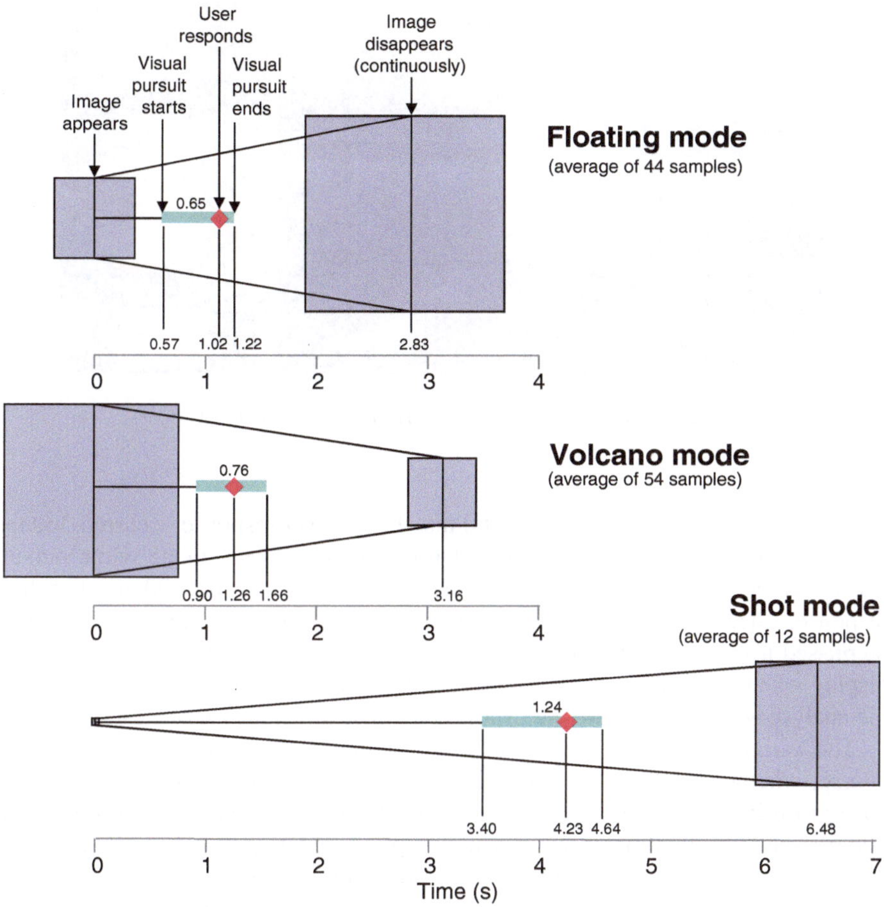

Floating mode
(average of 44 samples)

Image
appears

Visual
pursuit
starts

User
responds

Visual
pursuit
ends

Image
disappears
(continuously)

0.65

0.57   1.02 1.22   2.83

0   1   2   3   4

Volcano mode
(average of 54 samples)

0.76

0.90 1.26 1.66   3.16

0   1   2   3   4

Shot mode
(average of 12 samples)

1.24

3.40   4.23 4.64   6.48

0   1   2   3   4   5   6   7
Time (s)

Fig. 5.19   Summary of pursuit timing and image size data

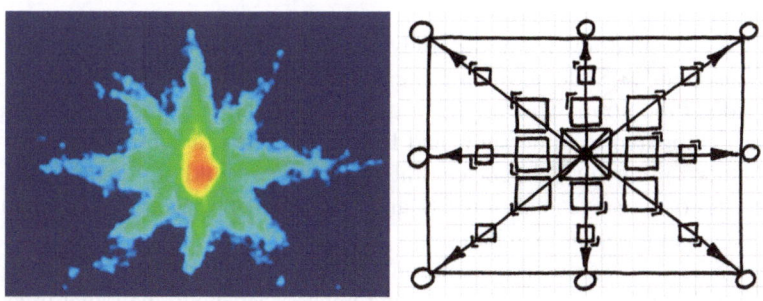

Fig. 5.20   Volcano mode cumulative heatmap (*left*, 5 trials), notation (*right*)

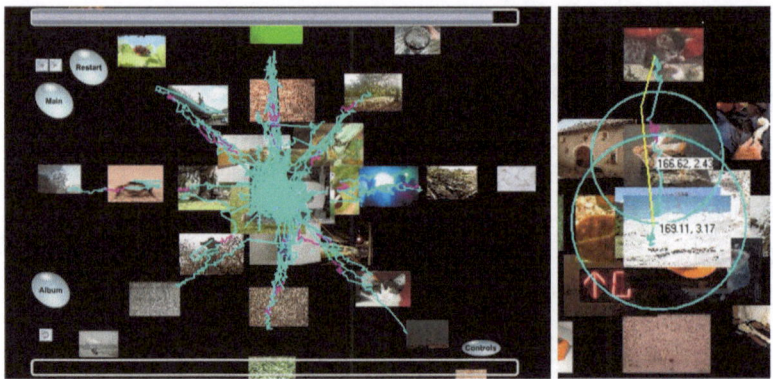

Fig. 5.21 Volcano mode, all tracks for one user (*left*), single pursuit, detail (*right*)

appearance of an image, and lasted 760 ms, though user response occurred 360 ms into the visual pursuit. The majority of user responses were made while pursuit was underway. On average, users continued to follow images for 400 ms after their response. There is, of course, a danger that closely spaced target examples may be missed if the user's attention is taken up following an existing target across the display when a second appears (there is at least one example of this occurring in the analysed data set). Figure 5.21 shows the tracking events for one user.

The variability of user behaviour is considerable. It is impossible to say, of course, whether a user is responding to a target image identified during steady gaze or during pursuit. In 3 cases out of the 54 the user responded before visual pursuit began. Sometimes users did not respond to a target and no gaze pursuit was recorded. At other times non-targets were followed but no response made: it is not clear if there was doubt in the user's mind or whether the image being followed was merely interesting.

5.3.3 Refining Volcano Mode

One of the clear findings from this study is that gaze behaviour for Volcano mode is not equivalent to Slide-show or Ring mode. Pursuit outside the central steady gaze region is the norm. This might be due to the relatively small size of the central image compared to Slide-show and Ring modes. It is a completely open question whether opting for a larger central image size would significantly alter the gaze process or target recognition rate and whether, if this were done, the moving radial image paths would then be used.

Figure 5.22 shows a possible conjectural redesign. The image size on appearance is made much larger, and the corresponding image sizes along the radial trajectories are larger also. Figure 5.19 indicates that the effective part of the display (i.e. where

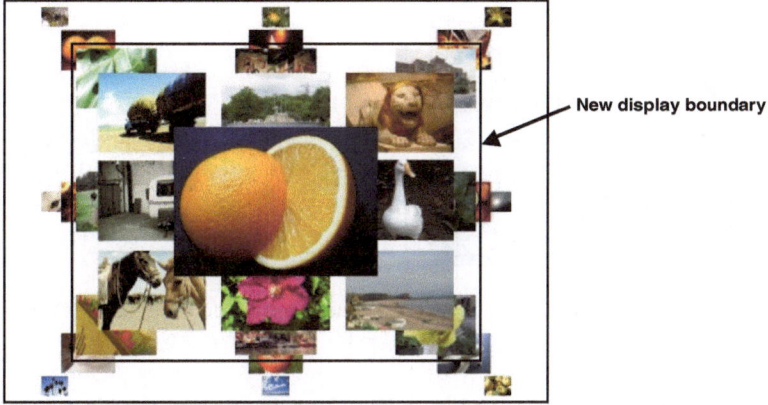

Fig. 5.22   Volcano mode, potential redesign (courtesy of T. Brinded)

recognition events have already occurred) is still well inside this new boundary, which would then be expanded to the edges of the display. Note, however, that the existing design is already quite effective from the point of view of target recognition (70 %, as reported by Corsato et al. 2008), and has a low fatigue rating (Fig. 5.17).

5.3.4 Floating Mode

Figure 5.23 (left) shows the cumulative heatmap for gaze activity for five trials of Floating mode, similar to that shown for Volcano mode (Fig. 5.20). The plots are apparently functionally similar—gaze is steady and concentrated at the centre where images appear (albeit smaller in this mode) and tracks outwards for potential targets (visual pursuit). It is interesting to note that the design changes between Volcano and Floating made little difference to the timings produced by the

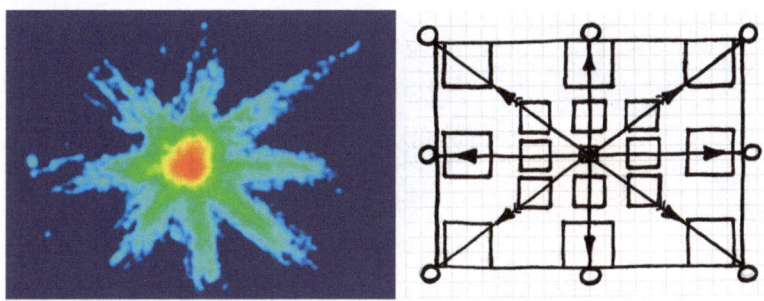

Fig. 5.23   Floating mode cumulative heatmap (5 trials) (*left*), notation (*right*)

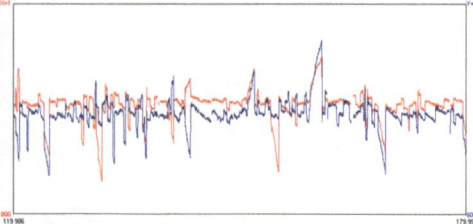

Fig. 5.24 Floating mode: selection of tracks from one user (60 s duration) (*left*), XY-T (*right*)

detailed analysis (summarised in Fig. 5.19). Gaze acquisition time is reduced, consistent with the reduced (and therefore less obscuring) size of the central image.

Figure 5.24 (left) shows a selection of tracks extracted from a 60 s segment taken from the complete 220 s sequence for one user. It may be seen that most of these pursuit tracks are directly associated with target recognition response—the magenta sections. Figure 5.24 (right) shows the XY-T plot for the 60 s extract. The larger visual pursuit movements are clearly visible, but the plot also indicates smaller tracking movements closer to the display centre.

5.3.5 Shot Mode

Corsato et al's Shot mode operates differently from Volcano and Floating modes. Images appear centrally and near the top of the display and move towards the bottom while getting larger. The cumulative heatmap (Fig. 5.25, left) for this mode shows that there is no gaze activity until the images have become a useable size.

Average pursuit start time for targets in this mode (Fig. 5.19) is 3.4 s and pursuit starts approximately halfway down the display (Fig. 5.25, left). Pursuit

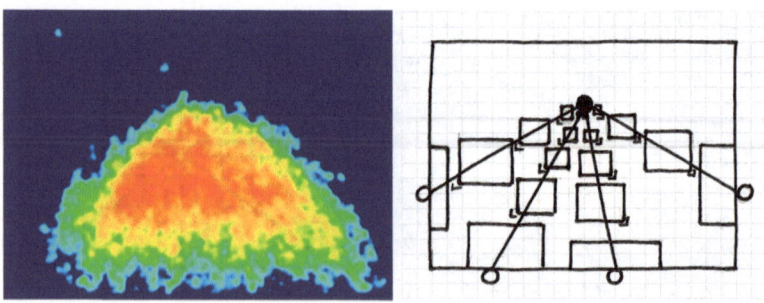

Fig. 5.25 Shot mode cumulative heatmap (5 trials) (*left*), notation (*right*)

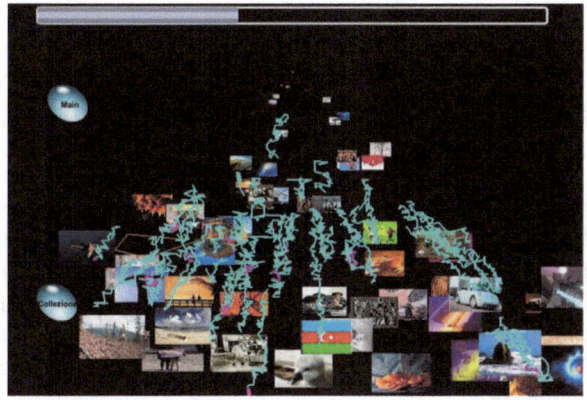

Fig. 5.26 Shot mode, all tracks for one user

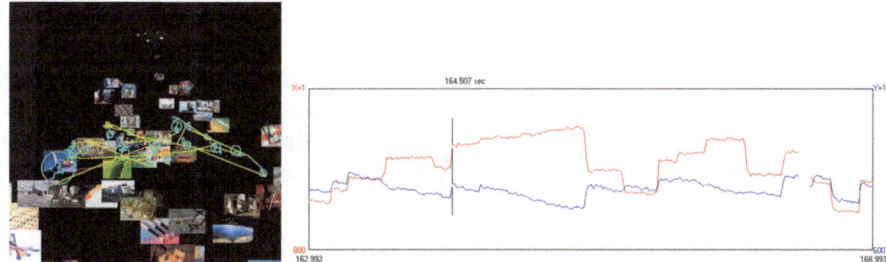

Fig. 5.27 Shot mode visual search behaviour detail (*left*), corresponding XY-T plot (*right*)

duration is also longer, on average, at 1.24 s. As with Volcano and Floating modes, there is clear visual pursuit behaviour along target image paths (Fig. 5.26), but in this mode there is also clear evidence of generally "horizontal" or side to side visual search between the pursuit events, rather than the steady gaze behaviour of Volcano and Floating modes.

Figure 5.27 (left) shows a six second detail of this horizontal side to side visual search behaviour. The user scans to and fro across the image area on the display, but generally stays within a relatively narrow horizontal band. Saccades range across the display width. Fixations are less stable than for a static image and tend to be somewhat elongated along the direction of the moving images. This six-second trace includes a single target pursuit event target identification (far left, large circle). Figure 5.27 (right) shows the XY-T plot for these 6 s. The single target pursuit event starts at 164.5 s (vertical line).

Each user appears to select a different image size range to search in this way, so that the central area of the cumulative heatmap appears spread vertically.

5.3.5.1 Refining Shot Mode

The analysis in the previous section indicates that half the display area in Shot mode is unused, and that the user is obliged to search the region of the display where images are at a minimum size for recognition. Altering the design so that it is reversed, with images appearing at their largest size at the bottom edge and tracking towards a point at the top of the display, would consistently provide a larger image for visual search while hardly compromising the user's ability to pursue the images into the display. A separate study (Mardell et al. 2009) indicates that gaze naturally searches close to the edge of continuous entry (apparently regardless of top, bottom, left or right) and follows features of interest into the display area. Whether these changes would improve recognition or user preference remains an open question.

5.3.6 Collage Mode

In Corsato et al. (2008) Collage mode[6] (illustrated previously in Fig. 3.12) images appear rapidly at random locations on the display and may be covered, completely or partially, at any time by a new image. The average time during which any given image was visible, either in its entirety, or partially obscured (see Sect. 3.3.3) was 6 s. However, the period during which images could be recognised was very variable, the shortest time for just over 300 ms, the longest for some 40 s.

The cumulative heatmap for this Collage mode is shown in Fig. 5.28. At first sight it might seem that the gaze pattern follows a conventional saccade and fixation visual search, apparently similar to that for Tile mode. However, in the more

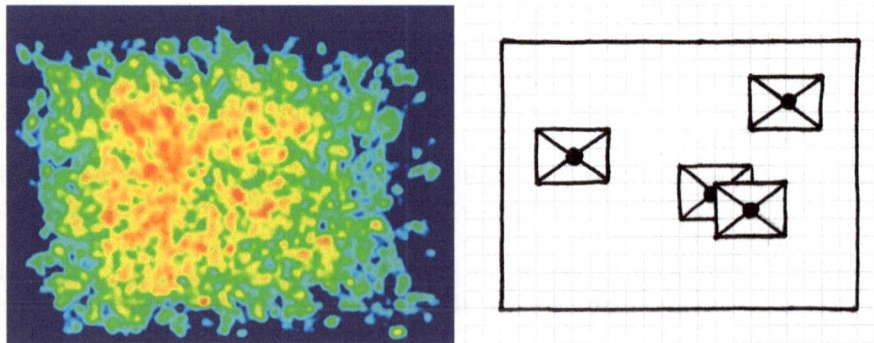

Fig. 5.28 Collage mode cumulative heatmap (5 trials), notation (*right*)

---

[6] This Collage mode RSVP differed from the design by Wittenburg et al. (1998, 2000, illustrated in Fig. 1.3) in that a very large number of images are visible on the display at any one time.

detailed analysis of 36 target appearances (from a single user), the user responded correctly on just 12 occasions. In all but three of these, target recognition could be interpreted as pre-attentive in that the targets were fixated on and responded to within one second. The average response time from gaze acquisition to key press was 702 ms.

The remaining recognition instances relied on a normal visual search pattern, which varied considerably in the time taken—the longest delay between appearance and reported recognition being 31.5 s! This would imply that pre-attentive recognition is very important to success in this example of Collage mode, and search less effective. Interestingly, although there is clear evidence from the detailed analysis that while the user fixated on at least 19 of the available targets, only 12 were positively identified. There was no instance of an identification response made without the user first fixating on the target image.

Collage mode is different from Tile mode in that pre-attentive recognition appears to be much more significant, yet there appears to be a conflict between searching and the potential interruptions posed by the continuous stream of new images. New images appear too quickly to be fixated on individually in turn such that the balance between search and immediate recognition must be maintained. Generally this example of Collage mode was neither successful in terms of effective recognition nor, with its extensive gaze travel, particularly popular with users.

5.3.7 Multiple Entry/Exit Modes, A Summary

Both Volcano and Floating modes showed a combination of steady gaze with interruptions due to visual pursuit events tracking potential targets across the display.

Shot mode also exhibited the visual pursuit of potential targets, but gaze showed a distinct tendency to a visual search strategy across the spread of images as they became large enough to identify. Shot mode was less effective and less popular than Volcano and Floating, showing a greater extent of gaze travel.

The gaze behaviour invoked by Collage mode, while superficially similar to the visual search strategy adopted in Tile mode, was both ineffective and unpopular—careful analysis showed a strong reliance on pre-attentive recognition, which was in turn often masked by the rapid appearance of distracting new images.

The most favoured and effective modes (Floating and Volcano) were also found to be associated with shorter gaze travel paths compared with the Shot and Collage modes. With regard to the number of correct images selected, Floating and Volcano scored highest, with Shot much lower. User preference was noticeably higher, and fatigue much lower, for Floating and Volcano modes compared with Shot and Collage modes. Thus, for all Corsato et al's measures of performance in these multiple entry/exit modes, Floating and Volcano were found to be superior to Shot and Collage.

5.4 Summary

This chapter has concentrated on the gaze responses of actual RSVP users. Previously, we identified four gaze strategies: visual search, steady gaze, nystagmus and visual pursuit, which, taken together, characterise our users' natural gaze responses to 10 very different RSVP designs.

In all, gaze analysis can give the RSVP designer significant insights into how a new design will be accepted by users. The designer can use eye gaze technology to refine designs to avoid pitfalls arising from perceptual artefacts.

References

Becker, W. (1991). Saccades. In R. H. S. Carpenter (Ed.), *Eye movements* (Vol. 8, pp. 95–137). Vision and visual dysfunction Boca Raton: CRC Press.

Cooper, K., de Bruijn, O., Spence, R., & Witkowski, M. (2006). A Comparison of static and moving presentation modes for image collections (pp. 381–388). *Proceedings of Advanced Visual Interfaces (AVI-2006)*.

Corsato, S., Mosconi, M., & Porta, M. (2008, May). An eye tracking approach to image search activities using rsvp display techniques, (pp. 416–420), *ACM, Proceedings of Workshop on Advanced Visual Interfaces, Naples*.

Mardell, J., Witkowski, M., & Spence, R. (2009). Detecting Search and rescue targets in moving aerial images using eye-gaze (pp. 67–70). *Proceedings of 5th International Conference on Communication by Gaze Interaction (COGAIN-09)*.

Wittenburg, K., Ali-Ahmad, W., LaLiberte, D., & Lanning, T. (1998). Rapid-fire image previews for information navigation (pp. 76–82). *ACM, Proceeding Conference on AVI*.

Wittenburg, K., Chiyoda, C., Heinrichs, M., & Lanning, T. (2000, January 26–28). Browsing through rapid-fire imaging: requirements and industry initiatives. In *Proceedings of Electronic Imaging '2000: Internet Imaging* (pp. 48–56). San Jose, CA, USA.

Chapter 6
Design

Abstract An interaction designer wishing to explore the potential of RSVP for a given application would expect to be provided with a concise set of guidelines requiring only a minimal understanding of the underpinning empirical evidence and theory. Both objective (e.g., image recognition) and subjective (e.g., fatigue) performances are of concern. Six questions typically asked by interaction designers concern those parameters that influence performance: they include the choice of visual style and the number of images simultaneously visible on the display. Following the presentation of design guidelines, examples of applications based on RSVP are provided.

keywords Interaction design • Design guidelines • Visual style • Presentation rate • Image size • Manual control • Simultaneous presentation of images • Applying RSVP

An interaction designer contemplating the use of RSVP in an application may wish to proceed directly to this chapter, but will consequently be unaware of the design considerations gradually accumulated in the earlier chapters. For this reason the present chapter will contain *brief* recollections of those considerations and the basis from which they were derived. The chapter then finishes with a discussion of a number of RSVP application designs as well as pointers to potentially useful directions of research.

6.1 User Tasks

An interaction designer minded to employ an RSVP mode in an application will have to consider not only the user tasks that require support but, in addition, user acceptance of the designed interface.

As pointed out in Chap. 1 it can be useful to consider two different tasks in which a user may be engaging, namely *exploration* and *search*. In some applications one or the other may dominate, but it is important to remember that even when the principal activity is search the user may, consciously or unconsciously, be forming a mental model of the information space encountered during that search. Indeed, in many situations, the visibility of context can be a most important consideration. Equally, to achieve their goal, a user may switch from exploration to search quite rapidly and frequently.

6.2 Issues in Design

For any task there will be two outcomes that the interaction designer will be striving for. One can be measured objectively, the other subjectively.

One is *task performance*. One example of measurable (objective) user performance is the accuracy of image identification in the course of undertaking an image recognition task. Another is the acquisition of the gist of a film. Yet a third is the time taken to locate a known image.

Subjective outcomes can include a user's perception of the fatigue they encounter in the execution of a task, their view as to whether the rate of presentation of images was too slow or too fast as well as their confidence regarding perceived task success. An outcome that is very difficult to predict is the extent to which an interface is visually attractive.

6.3 The Design Process

To develop guidance for the interaction designer we briefly reflect upon the design process that might be followed. In the very early, 'idea generation' stage of design the designer, with an intended application in mind, might well first sketch (Fig. 6.1a) the route to be followed by a collection of images, then tentatively indicate the approximate location of image frames/outlines on that route (Fig. 6.1b), perhaps also then adding (Fig. 6.1c) the—often fixed—location of appropriate text and other screen furniture associated with the application. Here, appropriate notation (see Chap. 3) indicates that image movement is continuous, and the image frame separation provides some indication of relative image speed at different points along the trajectory.

In the form shown in Fig. 6.1c the specification for the RSVP design is nevertheless incomplete. Lacking are two numerical parameters: one is the rate at which images appear (pace, p, images per second) and the other is the transit time (T_A) for any given image—in other words, the time for which an image remains visible. Perhaps tentatively, the designer might annotate the sketch, as in Fig. 6.1d, with values for these parameters (their choice is discussed in the following section). However, one detail may be added at this point. The designer may feel that, notwithstanding the lack of supportive experimental evidence mentioned in Chap. 3, the application would benefit if, at one location, each image is 'captured'. If that

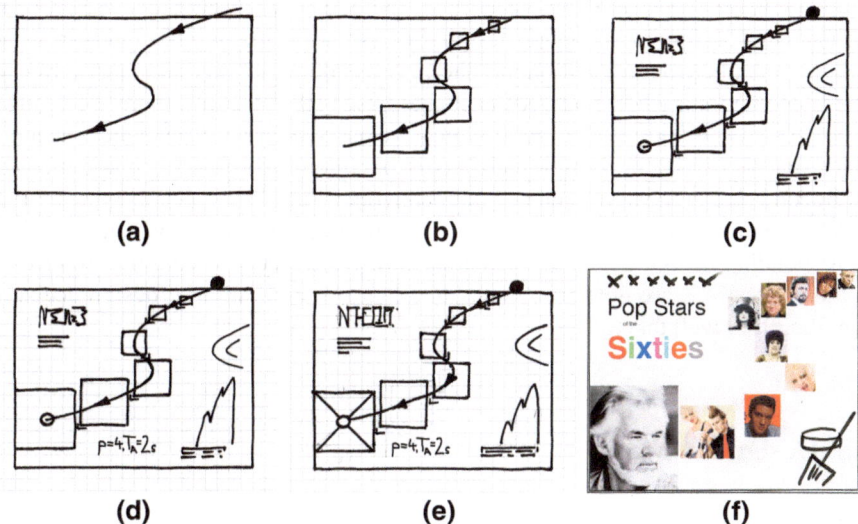

Fig. 6.1 A possible sequence of sketches created in the course of a first proposal for an RSVP application

is the case, the location and duration of the captured image might be as illustrated in Fig. 6.1e. The series of sketches might usefully lead to the creation of a paper prototype (Fig. 6.1f) to support formative evaluation by the designer as well as colleagues, client and representative users, and will often lead to design iterations. The approach illustrated in Fig. 6.1 can be taken in connection with any of the many modes introduced in this book.

6.4 Design Questions

The interaction designer will usually have five major questions in mind during the design activity:

- What is the most effective visual style (mode) for RSVP?
- How quickly can a collection of images be viewed?
- How large should the images be?
- Should presentation rate be fixed or user selected/controlled?
- How many images should simultaneously be visible on the screen?

Starting with the sort of questions a designer might typically ask and then proceeding to answers—as we do in this chapter[1]—may at first appear unusual. It can

---

[1] The approach adopted in this chapter is close to that described in Witkowski and Spence 2012.

be justified, however, in many ways. First, the questions listed above are arguably the principal factors in an RSVP design. Second, as in many other fields in which design typically proceeds with many factors considered in parallel, no step-by-step design methodology has emerged for image-based RSVP. Third, there are so many factors to jointly optimise, some of which are associated with aesthetic considerations, that flexibility in design decisions is paramount. In fact the guidelines that follow do not assume any sequence of design decisions.

We examine in turn each of the five questions listed above and establish corresponding design guidance, notwithstanding the fact that the topics referred to in those questions are not 'orthogonal' and that some iteration of design decisions may therefore well take place.

6.5 Design Answers

6.5.1 Visual Style

Especially for commercial applications a major consideration regarding visual design will be the degree to which an interface is found to be 'attention grabbing' and pleasurable. Such a property is difficult to define: even less does it lend itself to algorithmic description. Judgement of a design in this respect lies in the domain of an interaction designer's expertise, and is therefore not discussed further here. The interaction designer's expertise will undoubtedly benefit from the numerous books devoted to visual design (see, for example, Few 2009).

6.5.1.1 Static Mode

In some situations the answer to the question "What is the most effective visual style (i.e., mode) for RSVP?" may well be "a static mode", of which the Slide-show mode is the most commonly encountered in practice.

Common reactions to Slide-show mode point out the consequences of blinks (that can occur naturally at quite a rapid rate), and the fact that it can be quite stressful gazing at one point continuously for more than a few seconds, especially in the knowledge that each image is only visible for (say) 100 ms. For these reasons the Slide-show mode may only become of significant interest within a task such as that carried out by video editors when searching for an image in a collection of frames or a TV viewer engaging in fast-forward and rewind, especially where manual control of direction and speed are provided and when the visual display also represents context (Wittenburg et al. 2003b). At one extreme the provision of visual context (e.g., to help identify an approximate location) may justify a very high rate of presentation, while at the other the need to select exactly the right image may call for a very slow and manually controlled rate.

The Slide-show mode of RSVP may sometimes be relevant in circumstances closely related to the original 'book riffling' activity, as with the rapid presentation of key frames of a video in order to provide the gist of that video within a very short time. On TV channels, trailers of upcoming films are often presented at image rates not too far removed from ten per second.

6.5.1.2 Moving and Multiple Entry/Exit Modes

Moving RSVP modes are generally relevant to applications directed towards the consumer market, in sharp contrast to the professional market for which the (static) Slide-show mode may at times be appropriate. For consumer use heavy emphasis may well be placed upon the visual appeal of the application but, as explained above, it is not the function of this book to provide guidance in that respect. By contrast, firm guidance is provided immediately below that will apply to *any* RSVP mode involving image movement.

6.5.1.3 Gaze Travel

The discussion of eye-gaze behaviour in Chap. 5, and in particular the outcome of Corsato et al. (2008) experiment, strongly suggests that whatever trajectory is adopted by the interaction designer as the route to be taken by images, those images should *first* become capable of recognition at an essentially *single* location. That this criterion is satisfied for the Volcano mode shown in Fig. 6.2 (top) is suggested by the accompanying heatmap of gaze activity. It is not satisfied for the shot mode as demonstrated in Fig. 6.2 (lower).

6.5.1.4 Image Speed

The experimental evidence discussed in Chap. 5 suggests that image speed might usefully be kept comfortably below 1,000 pixels per second. A rough check of this requirement can easily be estimated from the sketch of Fig. 5.1d from knowledge of the total visibility time T_A of an image and the length of the image path.

6.5.1.5 Capture Frame

It is generally believed that the value of a moving RSVP mode can be enhanced if, at some fixed location, every image is captured for a short while, typically for 100–200 ms. As already mentioned, there is no experimental evidence to support this guidance: from what has been said in earlier chapters the interaction designer must judge, from knowledge of the application, whether a capture frame will enhance performance. A judgement in this respect might well take place during the first review of a prototype.

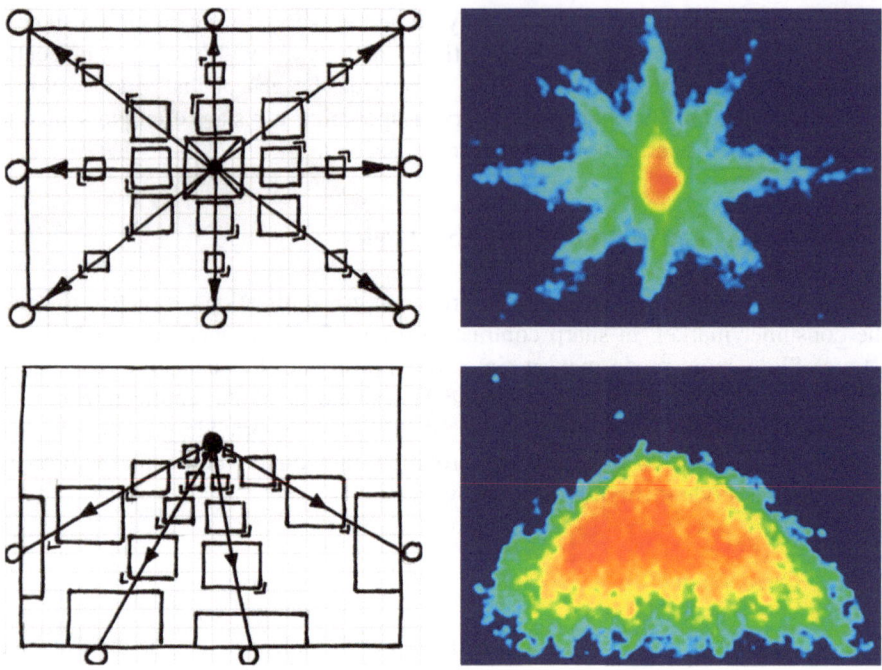

Fig. 6.2 Notational description and gaze heatmap for a volcano mode (*top*), notational description and gaze heatmap for a shot mode (*lower*)

6.5.2 Task Completion Time

"How quickly can a collection of images be viewed?" is a question motivated principally by the common need for a task to be completed within a reasonable time. Again, it is appropriate to answer this question separately for static and moving modes.

6.5.2.1 Static Mode

For Slide-show (static) mode the only parameter available for design is the pace, p—the rate at which new images appear on the display. Reference to Chap. 2 will confirm that a value of pace equal to ten per second should be regarded as a maximum, in which case the time taken for the collection of N images to be seen is at least N/10 s. However, as discussed above in Sect. 6.5.1.1, a task for which Slide-show mode is appropriate (for example, a video editor's exploration and search—both fast and slow—of a video sequence) would almost certainly benefit from an extensive range of pace under manual control.

6.5.2.2 Moving Modes

For moving modes a wider design flexibility is available, though a first consideration is simple: the higher the pace (p), the shorter in general will be the total image collection presentation time. An additional factor is the degree of overlap between successive images. The discussion of Chap. 3 showed that some degree of obscurement—even up to 50 % in some situations—might be tolerable for many types of image, and also identified the existence of a trade-off between pace and recognition task success. A decrease in image speed would increase overlap, but this disadvantage might be more than compensated by the additional time for which a given image is wholly or partially visible. The results discussed in Chap. 3 were derived for the Diagonal mode of RSVP, but would be expected to be generally applicable to other modes. In considering the use of overlap two considerations arise. First, to some extent, obscurement can be ameliorated by the existence of an unobscured capture frame. Second, if at some point in an overlapped sequence of images, each image 'rotates' to be visible in its entirety (as with CoverFlow—see Fig. 1.6) then, again, the overlap present elsewhere in the sequence may be less serious.

6.5.2.3 Multiple Entry/Exit Modes

All the considerations discussed above for moving modes apply to multiple entry/exit modes which, by their nature, offer additional potential for the confirmation of image recognition: there is more than one path that an image can follow before disappearing. This single feature may reduce considerably, or even eliminate, the need for image overlap.

6.5.3 Image Size

"How large should the images be?" is a difficult question to answer, especially if visual appeal is an important factor. Image features such as colour, distinctiveness, layout and previous exposure, as well as the manner in which a reduced size of image (a 'thumbnail') is generated from an original, all affect recognition. A factor to keep in mind is the fact that text, and especially its size, font and colour, constitutes an image: indeed, many products that might be advertised in floating mode, for example, may be instantly recognisable from familiar product name text (e.g., "GODIVA") rather than a product image (e.g., a box of chocolates). Experimental evidence relating recognition success to image *size* for representative images (including some containing meaningful text) is perhaps understandably hard to come by.

   In one sense the answer will become obvious during the evaluation of a design, and necessary adjustments will be made before it is finalised. Limited experimental evidence available from the literature (Kaasten et al. 2002; Suh et al. 2003;

Burton et al. 1995) broadly suggests that an image resolution of about 300×300 pixels will lead to high confidence of recognition, while a resolution of 100×100 might lead to around 60 % correct recognition.

6.5.4 Manual Control

The execution of a task using RSVP, whether that task involves exploration or search or a combination of the two, may demand, for its most efficient execution, a wide range of image visibility durations. In early RSVP applications speed and direction control involved a mouse-click on one of a number of buttons, as illustrated in Fig. 6.3.

However, as Wittenburg et al. (2000) found, a user, on noticing an image of interest, would typically then have to search for and click the stop button, by which time the image would have disappeared. A much improved method of control was to use mouse-over in place of click. With the Floating mode example the speed control (Fig. 6.4) was analogue: placement of the mouse-cursor towards the tips of the arrows increases the speed of presentation and towards the tails decreases the speed. Stopping the presentation is straightforward—the mouse cursor is moved away from the arrows. Mouse control *per se* is not a requirement: Brinded et al. (2011) describe a system in which a rotating knob controls either position or speed.

Evaluations carried out by Wittenburg et al. (2000) also suggested that different speeds might be associated with different tasks, with 'perusal' typically involving

Fig. 6.3 Discrete speed control (after Wittenburg et al. 2000)

Fig. 6.4 Analogue speed control (after Wittenburg et al. 2000)

a pace of about one image per second, 'scanning' involving one to two images per second and fast movement to a known region reaching four images per second. These were approximate observations, but give some idea of speeds that might be made available in an application.

6.5.5 Simultaneous Visibility of Images

The question "How many images should simultaneously be visible on the display" could equivalently, though approximately, be rephrased as "how long should each image remain visible?"

A characteristic feature of RSVP is the opportunity to recognise an image satisfying a specific criterion *pre-attentively*, and therefore very rapidly and without conscious cognitive effort. If for any reason that recognition is in doubt, and needs to be confirmed or rejected, then *attentive* revisitation must take place. Obviously, if infinite time is available for such attentive examination, all doubt can be removed. However, by the very nature of RSVP, the constant arrival of new images and the departure of old ones, some of which may satisfy the criterion, is such that very little time can be devoted to attention as opposed to pre-attention. There is therefore a case to be made for the limited retention of each image somewhere on the display, perhaps for a period lasting between one and two seconds, especially if this retention takes the form of predictable movement along predictable routes as in the Volcano mode (see Fig. 5.21). The previously discussed heatmaps for Volcano and Floating modes placed such revisitation in evidence, indicating that in such circumstances gaze almost always follows an image trajectory.

6.5.6 Summary of Design Considerations

We have identified, above, a number of design considerations; in fact, so many that they may be difficult to keep in mind. What may therefore be helpful to the interaction designer is a 'global' view of the major design considerations such as the one sketched in Fig. 6.5. As explained, pace (p) and image speed (s) are major factors in a design, and it is possible to represent some of the principal "soft" boundaries influencing a design in (p, s)-space. Together these boundaries define a central region that, at the very least, and tentatively, may be regarded as a "safe" location for the point (p, s) representing two major parameters of a design. Intentionally, numerical values are not assigned to the s-axis because the position of the various soft boundaries will depend significantly upon the RSVP mode being considered. The extent to which the boundaries may be approached or even violated is of course at the discretion of the interaction designer.

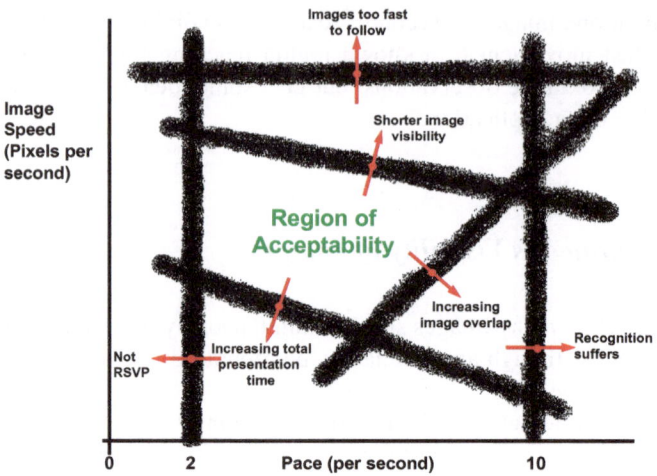

Fig. 6.5 A general summary of limitations affecting the design of an RSVP application

6.6 Evaluation

Evaluation of a design can occur at different stages in the development of an RSVP application.

A first evaluation might well occur at the stage at which a paper prototype has been constructed (e.g., Fig. 6.1f). Using this sketch, critical views can be elicited from professional colleagues, potential users and the client for whom the application is being developed, with the potential for useful design iteration. At this stage the satisfaction or otherwise of criteria regarding likely gaze travel and image speed can begin to be estimated.

When a first—and probably interactive—implementation is available an informal pilot experiment may be carried out to reveal any design weaknesses. Unless an experiment seeking statistical significance is planned, it would appear that 5–6 users would reveal most of the serious problems that need attention.[2] If eye-gaze detection is available this might provide a useful check on gaze behaviour. The many experiments described in this book identify ways in which both objective and subjective performance associated with a 'final' design can be observed.

6.7 Applying RSVP

The considerations that influence the design of an application exploiting RSVP can be numerous. To provide illustrations of this influence we briefly describe six case studies.

---

[2] See, for instance, http://www.measuringusability.com/blog/five-history.php.

6.7.1   Video Title Selection

Early in the commercial development of RSVP applications, Wittenburg et al.
(2000) investigated three generic applications: video title selection; window shopping; and people directories. The applications consequently ranged in content
from images *per se* (e.g., family photographs) to images constituting surrogate
content (e.g., video) and images acting as adjuncts (as in a product catalogue):
they are discussed respectively in this section, in Sect. 6.7.2 and then in Sect.
6.7.3. A major outcome was the design of improved manual speed and direction
controls, as discussed earlier in Sect. 6.5.4.

The *SeeHere* interface for entertainment video title selection, already illustrated
in simple form in Chap. 1, is illustrated more completely in Fig. 6.6. *SeeHere* is
seen in title skimming mode: the user has narrowed down a working set of video
titles—seen on the left—for further consideration, after which the dynamic
(collage-mode) presentation proceeds at a pace determined by the positioning of
a mouse over one of seven control buttons just below the image display. A click
on an image or title leads to additional text describing the video or to an option to
play a trailer.

6.7.2   Window Shopping

In an effort to introduce some elements of visual sensory engagement and enjoyment
into the activity of window shopping Wittenburg et al. (2000) invented the floating
mode of RSVP, introduced earlier and repeated in Fig. 6.7. Here the lesson learned
about speed and direction control with the *SeeHere* interface led to mouseover

Fig. 6.6   The SeeHere interface for video title selection (courtesy of Kent Wittenburg)

control: placement of the cursor towards the tips of the arrows increases the speed of presentation, and towards the tails decreases the speed. Placement away from the arrows in any direction causes the presentation to stop.

6.7.3 People Finding

The third RSVP application reported by Wittenburg et al. (2000) addressed the problem of browsing an organisational directory, and is illustrated by the sketch of Fig. 6.8. A user can employ it for conventional look-up, but can also rapidly browse the collection of faces. As a consequence it is particularly relevant to a situation in which a user cannot remember a person's name, location or organisation but would recognise their face: it exploits a human user's cognitive ability for rapid facial recognition. The authors point out that "Being able to attach faces to names would be a positive step in improving the basic social interactions necessary for developing effective business organisations".

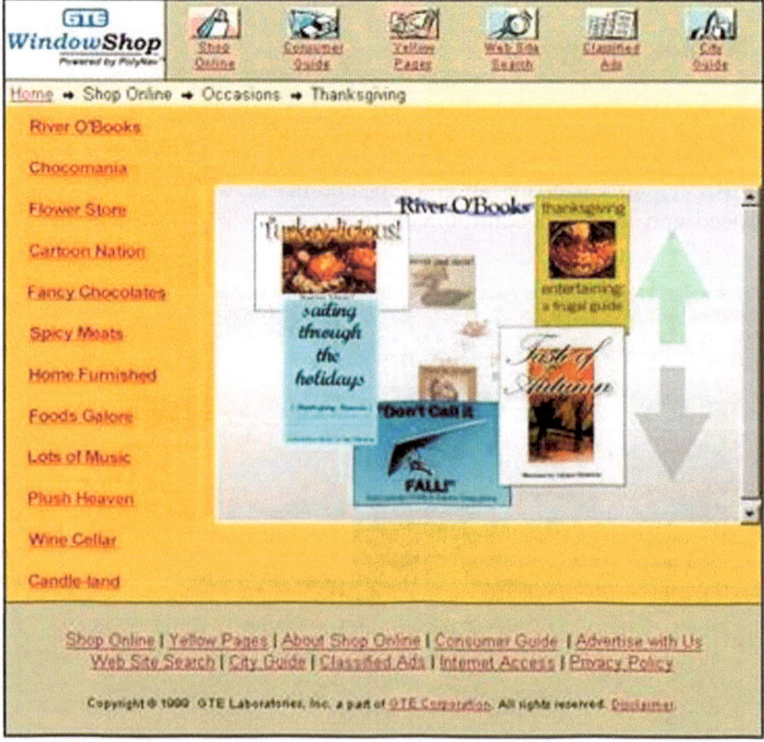

Fig. 6.7 The floating mode of RSVP, supporting window shopping (courtesy of Kent Wittenburg)

Fig. 6.8 An organisational directory (courtesy of Kent Wittenburg)

6.7.4 Video Fast-Forward and Rewind

A task that is repeatedly daily in millions of homes is that of fast-forwarding or rewinding a video or recorded television programme. The reasons are varied: to skip a commercial, to briefly check the weather during a programme currently being watched or to refer back to a particular scene. Fast forwarding and rewind is also undertaken by professional video editors who have the same requirements for an interface that will enable them to perform their task of composing a video production.

In conventional multi-media systems browsing takes place one frame at a time, and has many disadvantages. A commercial application of RSVP to support fast-forwarding and rewind was developed by Dr. Kent Wittenburg (a pioneer of RSVP applications) and his colleagues (Divakaran et al. 2005). To satisfy the needs of users their novel interface designs exhibit one essential characteristic—they preserve *temporal context* by displaying past and future frames at reduced size. One of many interfaces they explored during the development process is illustrated in Fig. 6.9: it is overlaid on top of a conventional full-screen fast-forward/rewind interface.

The so-called 'fisheye' layout in Fig. 6.9 shows the current frame (where playback will begin when selected) at full size in the centre: on each side, adjacent frames are both reduced in size and overlapped to an increasing degree. Even towards the end of the interface of Fig. 6.9 scene changes are immediately visible, and serve to support navigation. With the contextual information provided at the sides of the current frame consumers are better able to identify upcoming points of interest, scene boundaries and camera movement and, with familiar browsing

Fig. 6.9 Temporal context in video-editing mode (courtesy of Kent Wittenburg)

controls, are more quickly and accurately able to traverse to a desired location in the stream. A variant of the interface of Fig. 6.9 shows (Divakaran et al. 2005) enlarged versions of three images (current; 30 s 'ahead'; and 5 s behind) placed respectively in the centre, the extreme left and the extreme right with, in between, overlapped intervening images of slightly reduced size.

Evaluation of the interface of Fig. 6.9 as well as others embodying temporal context showed that subjects were significantly more accurate using such interfaces than when using a standard VCR interface. On average, for example, subjects resumed playback 25 % closer to an intended position (for example, the first frame after a set of commercials) when using the novel fast-forward scheme than when using the standard fast-forward (Wittenburg et al. 2003b). Of the many variants of interfaces designed to support fast-forward and rewind, and one which also displayed valuable temporal context, the 'V' design illustrated in Fig. 1.16 became a product.

6.7.5 Software Agents

In Sect. 1.4.13 we looked briefly at MAPPA (Witkowski et al. 2003), an e-commerce system incorporating both a manually controlled RSVP to inspect products, and a software agent to provide advice and guidance to users. The MAPPA system was evaluated extensively (Witkowski et al. 2001). Users were guided during a MAPPA session (lasting between 5 and 10 min) and encouraged to use the various features of the system, calling up the agent in both helper and advisor modes and using the manually controlled RSVP product display. The original analysis in Witkowski et al. (2001) was concerned primarily with how users interacted with the agent: here we have added some analysis of user interaction with the manually controlled RSVP based product display.

Only part of the overall time dedicated to product inspection was spent using the manual RSVP control. Figure 6.10 shows gaze behaviour during a single scan of the manually controlled product "shelf" lasting about 14 s in total. It may be seen to start with a single long fixation near the start of the controlling slider bar while the user positions the cursor over the control.[3] The user then draws the control tab from left to right at a somewhat uneven rate causing a corresponding

---

[3] Successive cursor positions are shown as small red 'x's here.

 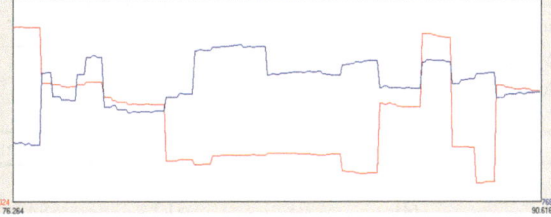

Fig. 6.10   Gaze behaviour during MAPPA RSVP activity

movement of the product display through about half of the 400 available products
(i.e. an *average* pace of about 13.7 images/s).

It is seen from the corresponding XY-T plot of Fig. 6.10 that gaze behaviour is
more analogous to Slide-show mode—long dwelling fixations tending to a steady
gaze strategy (average 1.3 s, longest 2.46 s) at various locations on the product
display. It is equally apparent that this user has managed the manual control so as
to avoid both the rapid nystagmatic gaze scanning that is associated with Stream
and Diagonal mode and the visual pursuit behaviours associated with the slower
moving image streams of multiple entry/exit modes such as Volcano and Floating.

6.7.6 Search and Rescue as RSVP

The Search and Rescue study of Mardell et al. (2012), previously described in
Sect. 1.4.14, serves as a useful illustration of the effects of changing from classic
visual search (wide area, long time) to RSVP (small area, short time). Recall that
Mardell et al. segmented a strip of aerial imagery into various degrees of segmentation and presented the segments at rates commensurate with their size to keep
the overall time for the complete presentation constant. Table 6.1 summarises the
parameters and resulting gaze travel speed results for their experiment. Embedded
amongst the images were a small number of hard to spot human "targets".
Examples are shown in Fig. 6.11 for two different segmentation degrees; note that
the targets that had to be identified and reported are to the same scale on the display.

Figure 6.12 illustrates the effects on gaze strategy as the presentation rate
increases from slow to rapid. Figures (1) show the full size presentation commensurate with a classical saccade and fixation visual search pattern (left, heatmap,
with the foveal area circled; right, a single example of a gaze track, with the target
circled). As the area is successively reduced and the rate increased (shown in
figures 2–6), a transition to a purely static gaze strategy is seen. The corresponding
reduction in gaze speed (column 7, Table 6.1), confirms the change.

The fact that the extent of gaze travel decreases as the degree of segmentation
increases is not surprising. By contrast, what is very surprising is the observation
that target recognition is substantially the same (around 65 % in the example shown)

Table 6.1 Search and Rescue parameters and corresponding gaze speeds for different segmentation degrees (reprinted from Mardell et al. 2012)

Segmentation degree	Number of tiles	Tile visibility time—(ms)	Image size, pixels		Subtended visual angle on display		Average gaze speed pix/s
			Width	Height	Horizontal	Vertical	
1	15	3,878	1,024	768	22.62°	17.06°	1,108
2	60	970	512	384	11.42°	8.58°	739
3	135	431	342	256	7.64°	5.72°	456
4	240	242	256	192	5.72°	4.30°	430
5	375	155	205	154	4.59°	3.45°	394
6	540	108	171	128	3.83°	2.86°	415

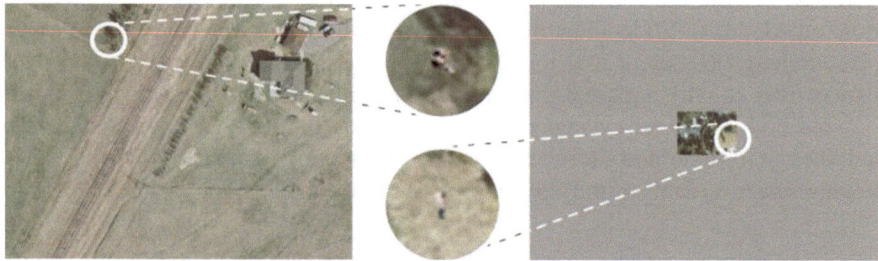

Fig. 6.11 Example targets embedded in the aerial imagery (courtesy of James Mardell)

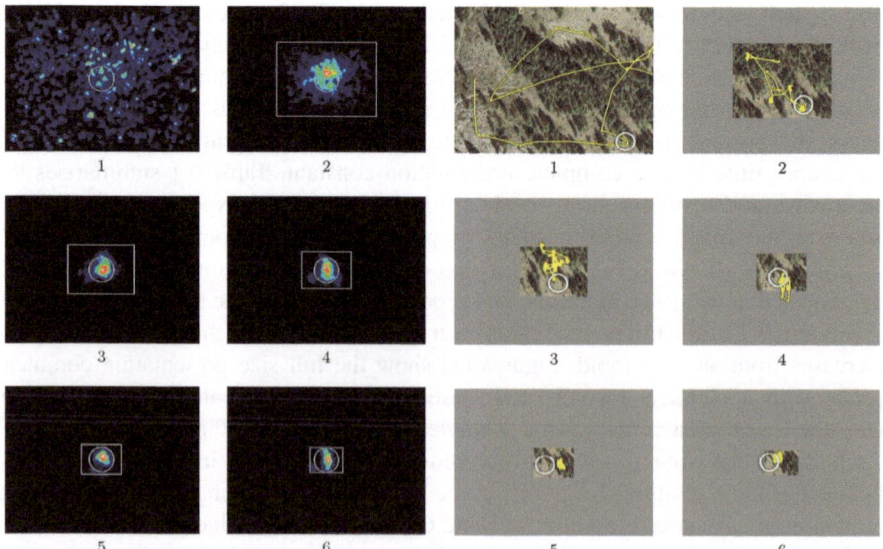

Fig. 6.12 Cumulative heatmaps for different segmentation levels (*left*) , gaze examples (*right*) (courtesy of James Mardell)

independently of the degree of segmentation. This result, together with the evidence shown in Fig. 6.12, suggests that the human visual processing system is remarkably robust in its recognition abilities across a wide range of presentation styles.

6.8 Potential Applications

The many modes and applications of RSVP discussed in this book inevitably lead to ideas concerning other possible fields of application, ideas that might be worth exploring. In this section we draw attention to potential applications that might benefit from further investigation.

6.8.1 Airline Baggage Inspection

The inspection of airline baggage to detect the presence of forbidden articles shares many of the challenges experienced in search and rescue; the consequence of a failure to detect a prohibited item could range from the serious to the catastrophic. Nevertheless the two applications differ in one major respect: with search and rescue the size of the target is so small that segmentation RSVP can safely be deployed, whereas the size of a target (e.g., knife, revolver) within a baggage can be substantial as a proportion to the whole, thereby introducing new challenges to the segmentation approach. Gale et al. (2000) investigated operator performance when identifying simulated Improvised Explosive Devices (IED) under rapid presentation conditions.

In airline baggage inspection the relevant images are mainly obtained by X-ray imaging (Fig. 6.13) and can be presented statically to a trained inspector. Currently, such a static image is examined for a few seconds by one or more trained staff,

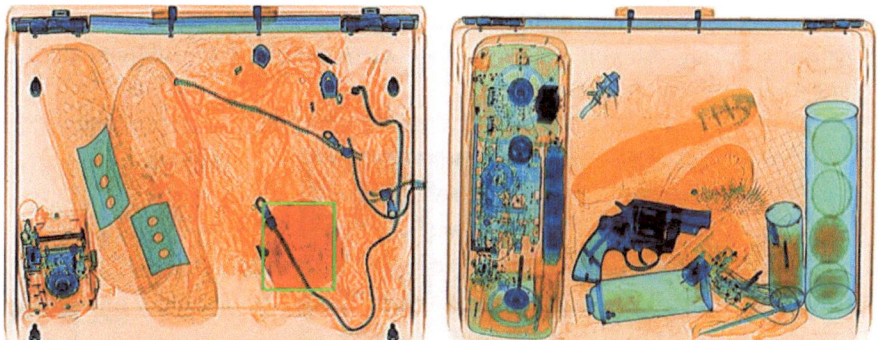

Fig. 6.13   X-ray images of airline baggage (courtesy of Smiths Detection [4])

---

[4] www.smithsdetection.com

but the potential of presenting the whole or part of such an image in Slide-show mode with a pace as low as 3 per second may be worth exploring. In this context it may be relevant to recall the findings of Kundel and Nodine (1975) who reported that trained radiologists were able to identify anomalies in lung X-rays with 70 % success when those X-rays were visible for only 200 ms each. Clearly, an issue is the use of more than one inspector. One factor in planning an approach to baggage inspection is acknowledgement of the very low prevalence of target occurrence, which has been shown to reduce identification success (Wolfe et al. 2005).

6.8.2 Blood Cells

The clinical investigation associated with many diseases involves the testing of a blood sample. In some situations a slide is prepared (Fig. 6.14a) containing a number of blood cells. It is possible, by using watershed segmentation (Roerdink and Meijster, 2001), to segment the individual cells (Fig. 6.14b) which can then (Fig. 6.14c) be treated as separate images. Clinical considerations may require the cells to be examined as a group in a single image, as in Fig. 6.14a and b. However, in some situations, as Forlines and Balakrishnan (2009) have pointed out, it may be appropriate to present the separate individual cell images to an investigator in Slide-show mode (as in Fig. 6.14c).

6.8.3 Crowd Screening

In certain security situations it may be necessary to examine the many faces within a large crowd to identify the presence or otherwise of specific individuals for whom facial images are already available and characterised by certain facial dimensions. One automatic technique available for this purpose is FaceAlert (www.facealert.com) and has been shown to be very effective. However, if there are circumstances where recognition must be carried out by a human being it may be worth exploring the potential advantage of presenting faces individually, for examination, in Slide-show mode.

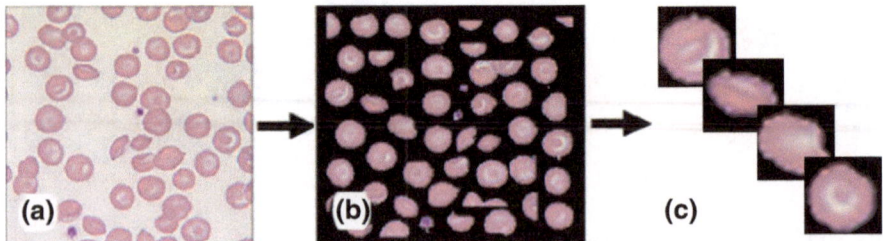

Fig. 6.14 **a** a blood cell slide, **b** the result of watershed segmentation, and **c** individual cells extracted for RSVP presentation (adapted from Forlines and Balakrishnan 2009)

6.9 Concluding Thoughts

From the earliest studies of pre-attentive processing, which established our surprising ability to perform certain tasks rapidly and without conscious cognitive effort, the technique of Rapid Serial Visual Presentation has developed along many lines. Advances in graphical processing, and computational power in general, have provided early innovators with the flexibility to explore a wide variety of modes (Chap. 1) in a range of practical applications. In parallel, advances in our understanding of the human visual processing system (Chap. 2) have helped to underpin those applications. Nevertheless, the range of possible modes (Chap. 3), as well as the manner of their detailed design, has emphasised the dilemma that the successful design of an RSVP application rests heavily upon the creative skill of the interaction designer. However, more recent studies of the eye-gaze behaviour associated with a number of modes (Chap. 4) has established empirical guidelines of value to the interaction designer. In particular, the need to be concerned with saccadic behaviour (Chap. 5) has emerged as a major guideline for the designer. It may not be an unjustifiable claim that the interaction designer contemplating the use of RSVP in an application is now provided with more useful guidance (Chapter 6) than was the case a decade ago.

RSVP is still being developed, even though some underlying empirical evidence dates back many decades. There are still many gaps in our knowledge, each of interest to researchers and designers in different disciplines. With regard to the underlying perceptual and cognitive processes there is still much to be learned, and some of the unexplained experimental results reported here will provide appropriate researchers with food for thought. There are also many unanswered questions in the mind of interaction designers who may consider the application of RSVP but need to be informed of user experience and, especially, acceptance of the many different features of an application exploiting RSVP. It will not be surprising if the creativity of those designers identifies useful new RSVP modes and, *inter alia*, new insights.

References

Brinded, T., Mardell, J., Witkowski, M., & Spence, R. (2011, July). The effects of image speed and overlap on image recognition (pp. 3–11). *Proceedings of 15th International Conference on Information Visualization (IV2011)*, London.

Burton, C.A., Johnston, L. J., & Sonenberg, E. A. (1995). Case study an empirical investigation of thumbnail recognition (pp. 115–121). *IEEE, Proceedings Conference on Information Visualization (INFOVIS' 95)*.

Corsato, S., Mosconi, M., & Porta, M. (2008, May). An eye tracking approach to image search activities using rsvp display techniques (pp. 416–420). *ACM, Proceedings of Workshop on Advanced Visual Interfaces*, Naples.

Divakaran, A., Forlines, C., Lanning, T., Shipman, S., & Wittenburg, K. (2005, January). Augmenting fast-forward and rewind for personal digital video recorders (pp. 43–44). *IEEE International Conference on Consumer Electronics (ICCE)*, Digest of Technical Papers, (IEEE Xplore, TR2004-136).

Few, S. (2009). *Now you see it.* Oakland: Analytics Press.

Forlines, C., & Balakrishnan, R. (2009). Improving visual search with image segmentation (pp. 1093–1102). *ACM Procedings of Computer Human Interaction (CHI-09)*, Boston, MA, USA.

Gale, G., Mugglestone, M. D., Purdey, K. J., & McClumpha, A. (2000). Is airport baggage inspection just another medical image? *Proceedings of SPIE, 3981*, 184.

Kaasten, S., Greenberg, S., & Edwards, C. (2002). How people recognize previously seen web pages from titles, URLs and thumbnails. In X. Faulkner, J. Finlay & F. Detienne (Eds.) People and computers XVI (pp. 247–265) (*Proceedings of Human Computer Interaction 2002*), BCS Conference series.

Kundel, H. L., & Nodine, C. F. (1975). Interpreting chest radiographs without visual search. *Radiology, 116*, 527–532.

Mardell, J., Witkowski, M., & Spence, R. (2012). An interface for visual inspection based on image segmentation (pp. 697–700). *Proceedings of Working Conference on Advanced Visual Interfaces (AVI-12)*, Capri Island (Naples), Italy: ACM.

Roerdink, J., & Meijster, A. (2001). The Watershed transform: definitions, algorithms and parallelization strategies. *Fundamenta Informaticae, 41*, 187–228.

Suh, B., Ling, H., Bederson, B.B., & Jacobs, D.W. (2003). Automatic thumbnail cropping and its effectiveness (p. 9). *Proceedings of User Interface and Software Technology Conference (UIST-03)*.

Witkowski, M., & Spence, R. (2012). Rapid serial visual presentation: An approach to design. *Information Visualization, 11*(4), 301–318. Published online April 25, 2012, doi:10.1177/1473871612439643.

Witkowski, M., Arafa, Y. & de Bruijn, O. (2001, March). Evaluating user reaction to character agent mediated displays using eye-tracking technology (pp. 79–87). *Proceedings AISB'01 Symposium on Information Agents for Electronic Commerce*.

Witkowski, M., Neville, B., & Pitt, J. (2003). Agent mediated retailing in the connected local community. *Interacting with Computers, 15*, 5–32.

Wittenburg, K., Chiyoda, C., Heinrichs, M., & Lanning, T. (2000, January 26–28). Browsing through rapid-fire imaging: requirements and industry initiatives (pp. 48–56). *Proceedings of Electronic Imaging '2000: Internet Imaging*, San Jose, CA, USA.

Wittenburg, K., Forlines, C., Lanning, T., Esenther, A., Harada, S., & Miyachi, T. (2003b). Rapid serial visual presentation techniques for consumer digital video devices (pp. 115–124). *ACM, Proceedings UIST'03*.

Wolfe, J. M., Horowitz, T. S., & Kenner, N. (2005). Rare items often missed in visual searches. *Nature, 435*, 439–440.

Bibliography

Anstis, S. M. (1974). A chart demonstrating variations in acuity with retinal position. *Vision Research, 14*, 589–592.

Becker, W. (1991). Saccades. In R. H. S. Carpenter (Ed.), *Eye movements* (Vol. 8, pp. 95–137). Vision and visual dysfunction Boca Raton: CRC Press.

Brinded, T., Mardell, J., Witkowski, M., & Spence, R. (2011, July). The effects of image speed and overlap on image recognition (pp. 3–11). *Proceedings of 15th International Conference on Information Visualization (IV2011)*, London.

Burton, C.A., Johnston, L. J., & Sonenberg, E. A. (1995). Case study an empirical investigation of thumbnail recognition (pp. 115–121). *IEEE, Proceedings Conference on Information Visualization (INFOVIS' 95)*.

Chahine, G., & Krekelberg, B. (2009). Cortical contributions to saccadic suppression *PlusOne, 4*(9) (online).

Coltheart, V. (Ed.). (1999). *Fleeting memories cognition of brief visual stimuli*. Cambridge, MA: MIT Press.

Cooper, K., de Bruijn, O., Spence, R., & Witkowski, M. (2006). A Comparison of static and moving presentation modes for image collections (pp. 381–388). *Proceedings of Advanced Visual Interfaces (AVI-2006)*.

Corsato, S., Mosconi, M., & Porta, M. (2008, May). An eye tracking approach to image search activities using RSVP display techniques (pp. 416–420). *ACM, Proceedings of Workshop on Advanced Visual Interfaces*, Naples.

de Bruijn, O., & Spence, R. (2002). *Patterns of eye gaze during rapid serial visual presentation: Proceedings of AVI-02* (p. 11).

de Bruijn, O., & Tong, C. H. (2003). M-RSVP: Mobile web browsing on a PDA. In E. O'Neill, P. Palanque, & P. Johnson (Eds.), *People and computers: Designing for society* (pp. 297–311). London: Springer.

Diamond, M. R., Ross, J., & Morrone, M. C. (2000). Extraretinal control of saccadic suppression. *The Journal of Neuroscience, 20*–9, 3449–3455.

Divakaran, A., Forlines, C., Lanning, T., Shipman, S., & Wittenburg, K. (2005, January). Augmenting fast-forward and rewind for personal digital video recorders (pp. 43–44). *IEEE International Conference on Consumer Electronics (ICCE)*, Digest of Technical Papers, (IEEE Xplore, TR2004-136).

Duchowski, A. T. (2003). *Eye tracking methodology: Theory and practice*. New York: Springer.

Erdmann, B., & Dodge, R. (1898). Psychologische Untersuchung über das Lesen auf experimenteller Grundlage. Niemeyer: Halle.

Findlay, J. M., & Gilchrist, I. D. (2003). *Active vision: The psychology of looking and seeing*. Oxford: Oxford University Press.

R. Spence and M. Witkowski, *Rapid Serial Visual Presentation*, SpringerBriefs in Computer Science, DOI: 10.1007/978-1-4471-5085-5, © The Author(s) 2013

Few, S. (2009). *Now you see it*. Analytics Press.

Forlines, C., & Balakrishnan, R. (2009). Improving visual search with image segmentation (pp. 1093–1102). *ACM Procedings of Computer Human Interaction (CHI-09)*, Boston, MA, USA.

Gale, G., Mugglestone, M. D., Purdey, K. J., & McClumpha, A. (2000). Is airport baggage inspection just another medical image? *Proceedings of SPIE, 3981*, 184.

Healey, C. G., Booth, K. S., & Enns, J. T. (1996). High-speed visual estimation using pre-attentive processing. *ACM Transactions on Human Computer Interaction, 3–2*, 107–135.

Holmquist, L. E. (1997). *Focus+Context visualization with flip-zooming and zoom browser. Exhibit, CHI'97*.

Intraub, H. (1980). Presentation rate and the representation of briefly glimpsed pictures in memory. *Journal of Experimental Psychology: Human Learning and Memory, 6*, 1–12.

Intraub, H. (1981). Rapid conceptual identification of sequentially presented pictures. *Journal of Experimental Psychology: Human Perception and Performance, 7*, 604–610.

Intraub, H. (1984). Conceptual masking—the effects of subsequent visual events on memory for pictures. *Journal of Experimental Psychology: Learning, Memory, and Cognition, 10*, 115–125.

Intraub, H. (1999). Understanding and remembering briefly glimpsed pictures: Implications for visual scanning and memory. In V. Coltheart (Ed.), *Fleeting memories: Cognition of brief visual stimuli*. Cambridge: MIT Press.

Itti, L., & Koch, C. (2001). Computational modelling of visual attention. *Nature Reviews, Neuroscience, 2*, 194–203.

Itti, L., Koch, C., & Niebur, E. (1998). A model of saliency-based visual attention for rapid scene analysis. *IEEE Trans Pattern Analysis and Machine Intelligence, 20*, 1273–1276.

Kaasten, S., Greenberg, S., & Edwards, C. (2002). How people recognize previously seen web pages from titles, URLs and thumbnails. In X. Faulkner, J. Finlay & F. Detienne (Eds.) People and computers XVI (pp. 247–265) (*Proceedings of Human Computer Interaction 2002*), BCS Conference series.

Kimron, L., Shapiro, K. L., & Luck, S. J. (1999). The attentional blink: A front-end mechanism for fleeting memories. In V. Coltheart (Ed.), *Fleeting memories: Cognition of brief visual stimuli*. Cambridge: MIT Press.

Kundel, H. L., & Nodine, C. F. (1975). Interpreting chest radiographs without visual search. *Radiology, 116*, 527–532.

Lam, K., & Spence, R. (1997). Image browsing: A space-time trade-off (pp. 611–612). *Proceedings INTERACT*. London: Chapman and Hall.

McConkie, G. W. (1983). Eye movements and perception during reading. In K. Rayner (Ed.), *Eye movements in reading* (pp. 65–96). New York: Academic Press.

McConkie, G. W., & Currie, C. B. (1996). Visual stability across saccades while viewing complex pictures. *Journal of Experimental Psychology: Human Perception and Performance, 22–3*, 563–581.

Mardell, J., Witkowski, M., & Spence, R. (2009). Detecting Search and rescue targets in moving aerial images using eye-gaze (pp. 67–70). *Proceedings of 5th International Conference on Communication by Gaze Interaction (COGAIN-09)*.

Mardell, J., Witkowski, M., & Spence, R. (2012). An interface for visual inspection based on image segmentation (pp. 697–700). *Proceedings of Working Conference on Advanced Visual Interfaces (AVI-12), Capri Island (Naples)*. Italy: ACM.

Nowell, L., Hetzler, E., & Tanasse, T. (2001). Change blindness in information visualization: A case study. *IEEE Proceedings of Information Visualization*.

Osterberg, G. (1935). Topography of the layer of rods and cones in the human retina. *Acta Ophthalmologica, 13*(6), 11–103.

Porta, M. (2006). Browsing large collections of images through unconventional visualization techniques (pp. 440–444). *ACM, Proceedings AVI*.

Porta, M. (2009). New visualization modes for effective image presentation. *International Journal of Image Graphics, 9–1*, 27–49.

Potter, M. C., Staub, A., Rado, J., & O'Connor, D. H. (2002). Recognition memory for briefly presented pictures: The time course of rapid forgetting. *Journal of Experimental Psychology-Human Perception and Performance, 28*, 1163–1175.

Potter, M. C., & Levy, E. I. (1969). Recognition memory for a rapid sequence of pictures. *Journal of Experimental Psychology, 81*, 10–15.

Potter, M. C. (1975). Meaning in visual search. *Science, 187*, 965–966.

Potter, M. C. (1976). Short-term conceptual memory for pictures. *Journal of Experimental Psychology-Human Learning and Memory, 2*, 509–522.

Potter, M. C. (1999). Understanding sentences and scenes: The role of conceptual short-term memory. In V. Coltheart (Ed.), *Fleeting memories: Cognition of brief visual stimuli.* Cambridge: MIT Press.

Raymond, J. E., Shapiro, K. L., & Arnell, K. M. (1992). Temporary suppression of visual processing in an rsvp task—an attentional blink. *Journal of Experimental Psychology: Human Perception and Performance, 18*, 849–860.

Rayner, K. (1998). Eye movements in reading and information processing: 20 years of research. *Psychological Bulletin, 124*(3), 372–422.

Rensink, R. A. (2000). The dynamic representation of scenes. *Visual Cognition, 7*, 17–42.

Rensink, R. A., O'Regan, J. K., & Clark, J. J. (2000). On the failure to detect changes in scenes across brief interruptions. *Visual Cognition, 7*(1–3), 127–145.

Rensink, R. A., O'Regan, J. K., & Clark, J. J. (1997). To see or not to see: the need for attention to perceive changes in scenes. *Psychological Science, 8*, 368–373.

Roerdink, J., & Meijster, A. (2001). The Watershed transform: definitions, algorithms and parallelization strategies. *Fundamenta Informaticae, 41*, 187–228.

Rubin, G. S., & Turano, K. (1992). Reading without saccadic eye movements. *Vision Research, 32–5*, 895–902.

Salvucci, D. D., & Goldberg, J. H. (2000). Identifying fixations and saccades in eye-tracking protocols (pp. 71–78). *Proceedings of the Eye Tracking Research and Applications Symposium.* New York: ACM Press.

Simons, D. J., & Levin, D. T. (1998). Failure to detect changes to people during a real-world interaction. *Psychonomic Bulletin and Review, 5–4*, 644–649.

Spence, R., & Apperley, M. D. (1982). Data base navigation: An office environment for the professional. *Behaviour and Information Technology, 1*(1), 43–54.

Spence R. (1998) A Content Explorer. Information Engineering Report 98/08. Department of Electrical and Electronic Engineering: Imperial College, London, 1998.

Spence, R. (2002). Rapid, serial and visual: A presentation technique with potential. *Information Visualization, 1*(1), 13–19.

Spence, B., Witkowski, M., Fawcett, C., Craft, B., & de Bruijn, O. (2004). Image presentation in space and time: errors, preferences and eye-gaze activity, (pp. 141–149). *ACM Proceedings of Workshop on Advanced Visual Interfaces (AVI-04).*

Spence, R. (2007). *Information visualization: Design for interaction.* Prentice Hall.

Stark, L. W., & Choi, Y. S. (1996). Experimental metaphysics: The scanpath as an epistemological mechanism. In W. H. Zangemeister et al. (Eds.), *Visual attention and cognition* (pp. 3–69). Amsterdam: Elsevier.

Suh, B., Ling, H., Bederson, B. B., & Jacobs, D. W. (2003). Automatic thumbnail cropping and its effectiveness (p. 9). *Proceedings of User Interface and Software Technology Conference (UIST-03).*

Sun, L., & Guimbretiere, F. (2005). Flipper: A new method for digital document navigation (pp. 2001–2004). *ACM Proceedings CHI'05* (Extended Abstracts).

Tatler, B. W., & Wade, N. J. (2003). On nystagmus, saccades and fixations. *Perception, 32*, 167–184.

Treisman, A. (1985). Pre-attentive processing in vision. *Computer Vision, Graphics and Image Processing, 31*, 156–177.

Treisman, A. (1991). Search similarity and integration of features between and within dimensions. *Journal of Experimental Psychology: Human Perception and Performance, 17*(3), 652–676.

Treisman, A., & Gormican, S. (1988). Feature analysis in early vision: evidence from search asymmetries. *Psychological Review, 95–1*, 15–48.

Tse, T., Marchionini, G., Ding, W., Slaughter, L., & Komlodi, A. (1998). Dynamic Key-frame presentation techniques for augmented video browsing (pp. 185–194). *ACM, Proceedings of Conference on AVI*.

Ware, C. (2004). *Information visualization: perception for design*. Amsterdam: Morgan Kaufmann.

Witkowski, M., Arafa, Y. & de Bruijn, O. (2001, March). Evaluating user reaction to character agent mediated displays using eye-tracking technology (pp. 79–87). *Proceedings AISB'01 Symposium on Information Agents for Electronic Commerce*.

Witkowski, M., Neville, B., & Pitt, J. (2003). Agent mediated retailing in the connected local community. *Interacting with Computers, 15*, 5–32.

Witkowski, M., & Spence, R. (2012). Rapid serial visual presentation: An approach to design. *Information Visualization*, Published online April 25, 2012, doi:10.1177/1473871612439643.

Witkowski, M., & Randell, D. A. (2007). A model of modes of attention and inattention for artificial perception. *Bioinspiration and Biomimetics, 2*, S94–S115.

Wittenburg, K., Ali-Ahmad, W., LaLiberte, D., & Lanning, T. (1998). Rapid-fire image previews for information navigation (pp. 76–82). *ACM, Proceedings of Conference on AVI*.

Wittenburg, K., Nicol, J., Paschetto, J., & Martin, C. (1999). Browsing with dynamic key frame collages in web-based entertainment video services (Vol. 2, pp. 913–918). *IEEE Proceedings of the International Conference on Multimedia Computing and Systems*.

Wittenburg, K., Chiyoda, C., Heinrichs, M., & Lanning, T. (2000, Jan 26–28). Browsing through rapid-fire imaging: Requirements and industry initiatives (pp. 48–56). *Proceedings of Electronic Imaging '2000: Internet Imaging*. San Jose, CA, USA.

Wittenburg, K., Forlines, C., Lanning, T., Esenther, A., Harada, S., & Miyachi, T. (2003b). Rapid serial visual presentation techniques for consumer digital video devices (pp. 115–124). *ACM, Proceedings Symposium on User Interface Software and Technology (UIST-03)*.

Wittenburg, K., Lanning, T., Forlines, C., & Esenther, A. (2003a, June). *Rapid serial visual presentation techniques for visualizing a third data dimension: Proceedings HCI International Conference*, Crete.

Wolfe, J. M., Horowitz, T. S., & Kenner, N. (2005). Rare items often missed in visual searches. *Nature, 435*, 439–440.

Yarbus, A.L. (1967). Eye movements and vision. New York: Plenum (Originally published in Russian 1962).

Index

R. Spence and M. Witkowski, *Rapid Serial Visual Presentation*, SpringerBriefs
in Computer Science, DOI: 10.1007/978-1-4471-5085-5, © The Author(s) 2013